RENEWALS: 691-4574

DATE DUE

FEB 2 6			
NOV 10			
OCT 0 5			
FEB 2 6			
APR 1 4			
DEC 1 2			
MAY 0 3			
OCT 2 3			
NOV 4			
OCT 1 5			
AUG 0 5 2009			

Demco, Inc. 38-293

Library of Congress Cataloging-in-Publication Data

Environmental epidemiology.

"Proceedings of the Symposium on Exposure
Measurement and Evaluation Methods for Epidemiology,
co-sponsored by the Health Effects Research
Laboratory of the United States Environmental
Protection Agency and the Division of Environmental
Chemistry of the American Chemical Society at the
190th National Meeting of the ACS in Chicago,
Illinois, September 8-13, 1985" — P. 1.
    Includes bibliographies.
    1. Health risk assessment — Congresses. 2. Poisons —
Dose-response relationship — Congresses. 3. Biological
monitoring — Congresses. 4. Epidemiology — Congresses.

I. Kopfler, Frederick C. II. Craun, Gunther F.
III. Symposium on Exposure Measurement and Evaluation
Methods for Epidemiology (1985 : Chicago, Ill.)
IV. Health Effects Research Laboratory (Cincinnati,
Ohio) V. American Chemical Society. Division of
Environmental Chemistry. IV. American Chemical
Society. Meeting (190th : 1985 : Chicago, Ill.)
[DNLM: 1. Environmental Exposure — congresses.
2. Environmental Monitoring — congresses. 3. Environ-
mental Pollutants — analysis — congresses.
WA 671 E606 1985]
RA566.27.E58   1986   614.4   86-21316
ISBN 0-87371-073-8

LEWIS PUBLISHERS, INC.
121 South Main Street, P.O. Drawer 519, Chelsea, Michigan 48118

PRINTED IN THE UNITED STATES OF AMERICA

This book is dedicated to
Leland J. McCabe,
whose investigations of the
relationship of illness and disease to waterborne
contaminants during his 35-year federal career
serve as a cornerstone of many of our
current drinking water and bathing-beach water standards.
We are privileged to have worked with him
during the last 15 of these years.
He is both friend and mentor.

PREFACE

The epidemiologic approach is a valuable methodology for assessing the association of chemical exposure and occurrence of a disease in a human population, and should be used to supplement data obtained from clinical and toxicologic research. Epidemiology studies are important in the regulatory process because the results are necessary to elucidate the risk of human chemical exposure incurred by human beings without the uncertainty of interspecies extrapolation. Because the United States Environmental Protection Agency (U.S. EPA) is required to develop regulations under six separate legislative acts, these studies are usually conducted to provide information for estimation of risk of exposure through a given route or from a specific source.

Case control and cohort studies can provide a quantitative estimate of health risk association with various environmental exposures, but it is often difficult to assess relevant exposures for individuals because retrospective epidemiologic studies require estimates of past exposure which must be made in light of current information. While it is important to fully understand the sources, routes, and extent of exposure of individuals to environmental toxicants, obtaining such knowledge about the population included in an epidemiology study may not be practical and/or achievable in all cases. Prospective studies can be designed in which exposure measurements are included as part of the study, but these studies are generally not feasible because of the high costs of following a cohort for a long period to determine associations between exposure and an observed health effect.

It is important that the exposure data collected for epidemiologic studies be relevant and appropriate for both the study design and regulatory needs. The accurate measure or assessment of exposure is paramount because random misclassification of exposure for study participants can only bias the outcome of the study toward one or no association between exposure and disease. Most epidemiologic studies have assumed that exposure to a contaminant is an adequate surrogate for the dose. A major limitation of past studies has been a lack of information on dose, e.g., the amount of the contaminant or metabolite in body tissue or the amount that interacts with the target organ or tissue. Biologic markers of cumulative dose would also assist in improving the sensitivity of epidemiologic studies, and should be considered, whenever possible, to supplement the data collected on exposure to environmental contaminants.

This book contains papers presented at a symposium co-sponsored by the Health Effects Research Laboratory of the U.S. EPA and the Division of Environmental Chemistry of the American Chemical Society at the 190th National Meeting of the ACS in Chicago, Illinois, September 8–13, 1985. It brings together the thoughts and work of epidemiologists, chemists, and mathematical modelers. By gaining insight into each other's needs and capabilities, scientists can plan research that will allow the exposure of participants in epidemiologic studies to be more accurately assessed. It is important that chemists, biochemists, and toxicologists fully understand the capabilities and limitations of epidemiologic studies so that improved interdisciplinary studies can be conducted.

Frederick C. Kopfler
Gunther F. Craun

ACKNOWLEDGMENTS

We wish to thank the Division of Environmental Chemistry and the U.S. EPA for the financial support required to provide the excellent facilities for the symposium and to help defray the costs of travel and registration for many of the nonchemist and foreign speakers.

Each paper included in this proceedings has been critically reviewed by at least two peers. We wish to express our appreciation to the following reviewers: Julian Andelman, John Bosch, Emile Coleman, Charlotte Cottrill, Donna Cragle, Peter Farmer, Peter Gann, Roger Giese, Daniel Greathouse, Jack Griffith, Robert Herrick, Norman Kowal, Pasquale Lombardo, Charles Mann, Gary Marsh, Patricia Murphy, Michael Pereira, Paul Skipper, John Stanley, Lance Wallace, and Elaine Zeighami. The extensive review to ensure high quality and logical validity resulted in improvement of most of the papers. However, each paper must ultimately stand on its own merit. Because of the review process, not all presented papers were accepted for publication.

We also wish to thank Andrea Donoghue, Betsy Kostic, Brana Lobel, and Kate Schalk of Eastern Research Group for their invaluable assistance in making travel arrangements, assembling the manuscripts, coordinating the peer reviews, and editing and producing the camera-ready copy for this book.

REVIEW/DISCLAIMER STATEMENT

**FREDERICK C. KOPFLER** is presently Chief of the Chemical and Statistical Support Branch, Toxicology and Microbiology Division, Health Effects Research Laboratory, U.S. Environmental Protection Agency, Cincinnati, Ohio. He obtained his BS in chemistry from Southeastern Louisiana University in 1960 and advanced degrees in biochemistry and food science from Louisiana State University. After completing a National Research Council-sponsored postdoctoral appointment in the Pioneering Research Laboratory for Animal Proteins at the U.S. Department of Agriculture, he was involved in environmental research with the U.S. Public Health Service and, since its formation in 1970, has been associated with the U.S. Environmental Protection Agency (EPA). Dr. Kopfler has worked closely with epidemiologists in designing studies of the relationship between drinking water chlorination practices and cancer incidence in consumers, and studies of mineral and trace element content of drinking water and the occurrence of cardiovascular disease. His current research areas include isolation of organic contaminants from water for toxicological testing and the identification of the reaction products of chlorine with biological chemicals.

 **GUNTHER F. CRAUN** has served in various capacities over the past twenty years as an environmental engineer and epidemiologist with the U.S. Public Health Service and the U.S. Environmental Protection Agency (EPA). Since 1970, he has been associated with EPA's drinking water and health research activities. His current research interests include relationships between drinking water contaminants and cardiovascular disease, cancer, and infectious diseases. He received his education in civil engineering (BS) and sanitary engineering (MS) at Virginia Polytechnic Institute, and public health (MPH) and epidemiology (MS) at Harvard University. He is registered as a professional engineer in the Commonwealth of Virginia.

Mr. Craun has authored and coauthored numerous articles in the international scientific, public health, and engineering literature. The American Water Works Association and the New England Water Works Association have recognized Mr. Craun for his work on waterborne disease outbreaks and trace metals in the drinking water of the Boston metropolitan area. The EPA awarded Mr. Craun a meritorious performance citation for his participation in the Community Water Supply Study, which identified deficiencies in the nation's public water supplies.

Mr. Craun is a member of the International Association of Milk, Food, and Environmental Sanitarians Committee on Communicable Diseases Affecting Man, and from 1977 to 1982 he served as chairman of the American Water Works Association Committee on the Status of Waterborne Disease Outbreaks in the United States and Canada. Mr. Craun served as liaison representative to the National Research Council's Safe Drinking Water Committee from 1974 to 1977, and as a member of the World Health Organization Working Group on Sodium Chloride, and Conductivity in Drinking Water in 1978. He has also served as a member of the Water Pollution Control Federation Research Committee and the International Association on Water Pollution Research Study Group on Virology.

Mr. Craun is currently Coordinator of the Environmental Epidemiology Program in EPA's Health Effects Research Laboratory, Cincinnati, Ohio. In his present capacity, he works with a number of other government agencies, including the National Cancer Institute and Oak Ridge National Laboratory, on epidemiological studies of drinking water contaminants. He is also involved in projects with the National Academy of Sciences and the University of Pittsburgh Center for Environmental Epidemiology to identify new research areas and methodologies for environmental epidemiology.

CONTRIBUTORS

Julian B. Andelman, Graduate School of Public Health, University of Pittsburgh, Pittsburgh, PA 15261.

David W. Armentrout, PEI Associates, Inc., 11499 Chester Road, Cincinnati, OH 45246.

Herman Autrup, Laboratory of Environmental Carcinogenesis, Fibiger Institute, 70 Ndr. Frihavnsgade, DK-2100, Copenhagen, Denmark.

Eric Bailey, MRC Toxicology Unit, Medical Research Council Laboratories, Woodmansterne Road, Carshalton, Surrey, SM5 4EF, England.

Margot Barnett, Graduate School of Public Health, University of Pittsburgh, Pittsburgh, PA 15261.

Alfred M. Bernard, Faculty of Medicine, Unit of Industrial and Medical Toxicology, Catholic University of Louvain, Clos Chapelle-Aux-Champs, BP 30.54, B-1200, Brussels, Belgium.

Kathy E. Boggess, Midwest Research Institute, 425 Volker Boulevard, Kansas City, MO 64110.

Joseph J. Breen, Office of Toxic Substances, U.S. Environmental Protection Agency, 401 M Street, S.W., Washington, DC 20460.

Matthew S. Bryant, Department of Applied Biological Sciences, Massachusetts Institute of Technology, Cambridge, MA 02139.

Richard J. Caplan, Department of Biostatistics, Graduate School of Public Health, University of Pittsburgh, Pittsburgh, PA 15261.

Joseph Carra, Office of Toxic Substances, U.S. Environmental Protection Agency, 401 M Street, S.W., Washington, DC 20460.

M. Virginia Cone, Science Applications International Corporation, 300 South Tulane Avenue, Oak Ridge, TN 37830.

Charlotte A. Cottrill, Epidemiology Section, Health Effects Research Laboratory, U.S. Environmental Protection Agency, 26 West St. Clair Street, Cincinnati, OH 45268.

Amy Couch, Graduate School of Public Health, University of Pittsburgh, Pittsburgh, PA 15261.

Gunther F. Craun, Epidemiology Section, Health Effects Research Laboratory, U.S. Environmental Protection Agency, 26 West St. Clair Street, Cincinnati, OH 45268.

Lars O. Dragsted, Laboratory of Environmental Carcinogenesis, Fibiger Institute, 70 Ndr. Frihavnsgade, DK-2100, Copenhagen, Denmark.

Larry J. Elliott, Industrial Hygiene Section, Industrywide Studies Branch,

Division of Surveillance, Hazard Evaluations and Field Studies, National Institute for Occupational Safety and Health, 4676 Columbia Parkway, Cincinnati, OH 45226.

Peter B. Farmer, MRC Toxicology Unit, Medical Research Council Laboratories, Woodmansterne Road, Carshalton, Surrey, SM5 4EF, England.

Marialice Ferguson, Science Applications International Corporation, 300 South Tulane Avenue, Oak Ridge, TN 37830.

Peter Gann, Department of Family and Community Medicine, University of Massachusetts Medical School, 55 Lake Avenue North, Worcester, MA 01605.

Roger W. Giese, Department of Medicinal Chemistry, College of Pharmacy and Allied Health Professions, Northeastern University, 360 Huntington Avenue, Boston, MA 02115.

John E. Going, Midwest Research Institute, 425 Volker Boulevard, Kansas City, MO 64110.

Daniel Greathouse, Hazardous Waste Engineering Research Laboratory, U.S. Environmental Protection Agency, 26 West St. Clair Street, Cincinnati, OH 45268.

Anna S. Hammons,, Science Applications International Corporation, 300 South Tulane Avenue, Oak Ridge, TN 37830.

Ty D. Hartwell, Research Triangle Institute, P.O. Box 12194, Research Triangle Park, NC 27709.

Robert F. Herrick, Industrial Hygiene Section, Industrywide Studies Branch, Division of Surveillance, Hazard Evaluations and Field Studies, National Institute for Occupational Safety and Health, 4676 Columbia Parkway, Cincinnati, OH 45226.

Frederick C. Kopfler, Health Effects Research Laboratory, U.S. Environmental Protection Agency, 26 West St. Clair Street, Cincinnati, OH 45268.

Herman Kraybill, National Cancer Institute (retired), 17708 Lafayette Drive, Olney, MD 20832.

Frederick W. Kutz, Office of Toxic Substances, U.S. Environmental Protection Agency, 401 M Street, S.W., Washington, DC 20460.

Robert R. Lauwerys, Faculty of Medicine, Unit of Industrial and Medical Toxicology, Catholic University of Louvain, Clos Chapelle-Aux-Champs, BP 30.54, B-1200, Brussels, Belgium.

Charles E. Lawrence, Laboratory of Statistics and Computer Sciences, Wadsworth Center for Laboratories and Research, New York State Department of Health, Room C-323, Albany, NY 12201.

Pasquale Lombardo, Division of Chemical Technology, Center for Food Safety and Applied Nutrition, Food and Drug Administration, 200 C Street, S.W., Washington, DC 20204.

Gregory A. Mack, Battelle Columbus Laboratories, 505 King Avenue, Columbus, OH 43201.

Gary M. Marsh, Center for Environmental Epidemiology, A416 Crabtree Hall, Graduate School of Public Health, University of Pittsburgh, Pittsburgh, PA 15261.

Glenn J. Martin, Health Care Financing Administration, Bureau of Data Management and Strategy, Office of Information Resources Management, G-A-2 Meadows East Building, 6325 Security Boulevard, Baltimore, MD 21207.

Edo D. Pellizzari, Research Triangle Institute, P.O. Box 12194, Research Triangle Park, NC 27709.

C. Donald Powers, Science Applications International Corporation, 300 South Tulane Avenue, Oak Ridge, TN 37830.

Janet C. Remmers, Office of Toxic Substances, U.S. Environmental Protection Agency, 401 M Street, S.W., Washington, DC 20460.

Philip E. Robinson, Office of Toxic Substances, U.S. Environmental Protection Agency (TS-798), 401 M Street, S.W., Washington, DC 20460.

Linda S. Sheldon, Research Triangle Institute, P.O. Box 12194, Research Triangle Park, NC 27709.

David E. G. Shuker, MRC Toxicology Unit, Medical Research Council Laboratories, Woodmansterne Road, Carshalton, Surrey, SM5 4EF, England.

Paul L. Skipper, Department of Applied Biological Sciences, Room 56-313, Massachusetts Institute of Technology, Cambridge, MA 02139.

Charles M. Sparacino, Research Triangle Institute, P.O. Box 12194, Research Triangle Park, NC 27709.

John S. Stanley, Midwest Research Institute, 425 Volker Boulevard, Kansas City, MO 64110.

Cindy R. Stroup, Office of Toxic Substances, U.S. Environmental Protection Agency (TS-798), 401 M Street, S.W., Washington, DC 20460.

Steven R. Tannebaum, Department of Applied Biological Sciences, Massachusetts Institute of Technology, Cambridge, MA 02139.

Philip R. Taylor, Cancer Prevention Studies Branch, Division of Cancer Prevention and Control, National Cancer Institute, Blair Building, Bethesda, MD 20892.

William W. Thurston, Graduate School of Public Health, University of Pittsburgh, Pittsburgh, PA 15261.

Johnston Wakhisi, Department of Surgery, University of Nairobi, Nairobi, Kenya.

Lance Wallace, U.S. Environmental Protection Agency (RD-680), 401 M Street, S.W., Washington, DC 20460.

Kaiwen K. Wang, Office of Drinking Water (WH-550), U.S. Environmental Protection Agency, 401 M Street, S.W., Washington, DC 20460.

Nancy W. Wentworth, Office of Research and Development (RD-680), U.S. Environmental Protection Agency, 401 M Street, S.W., Washington, DC 20460.

James J. Westrick, Technical Support Division (ODW), U.S. Environmental Protection Agency, 26 West St. Clair Street, Cincinnati, OH 45268.

Elaine A. Zeighami, Health Effects and Epidemiology Group, Oak Ridge National Laboratory, Building 4500-S, F-256, Oak Ridge, TN 37831.

Harvey Zelon, Research Triangle Institute, P.O. Box 12194, Research Triangle Park, NC 27709.

CONTENTS

**SECTION IV: ASSESSMENT OF EXPOSURE TO
ENVIRONMENTAL CONTAMINANTS FOR
EPIDEMIOLOGIC STUDIES**

PART ONE: AIR EXPOSURES

PART TWO: WATER AND OCCUPATIONAL EXPOSURES

CHAPTER 1

DETECTION OF AFLATOXIN B_1 GUANINE ADDUCTS IN HUMAN URINE SAMPLES FROM KENYA

Lars O. Dragsted, Johnston Wakhisi, and Herman Autrup

INTRODUCTION

Aflatoxin B_1 (AFB) is a pentacyclic secondary metabolite produced by the molds *Aspergillus flavus* and *A. parasiticus*. The formation of AFB is particularly favored when the molds grow on cereals rich in starch in hot and humid environments [1]. These conditions are prevalent during storage of cereals in many tropical and subtropical countries with non-industrialized agriculture.

AFB is a liver carcinogen in several animal species, including trout, rodents, pigs, and non-human primates [2] and is a suspected human liver carcinogen [3]. In Kenya, Peers and Linsell [4] found good correlation between crude liver cancer rates and dietary intake of aflatoxins, as measured in cooking pot samples.

Human exposure to AFB has been demonstrated by determination of AFB or metabolites in viscera or body fluids. In urine samples from Gambia, levels of more than 100 ng/ml were found by enzyme-linked immunosorbent assay (ELISA) [5]. Levels of 0.0-0.6 g/ml AFB have been detected in human blood samples from liver cancer patients in Nigeria by simple chromatographic and fluorescence techniques [6,7]. High performance liquid chromatography (HPLC) or thin layer chromatography (TLC) have been used to detect similar or higher levels in samples of blood, urine, or viscera from Sudanese [8], Thai [9], or American children [10] with suspected AFB-related syndromes. In the latter study, AFB was also demonstrated at similar levels in samples from healthy controls. Low levels of AFB (20-56 pg/ml) have been demonstrated in blood samples from healthy Japanese by a combination of radioimmunoassay (RIA) and chromatographic techniques, followed by a mass spectrometric verification [11]. In a pilot study with determination of AFB by RIA in American urine samples, levels of 5-50 pg/ml were found [12]. Immunoreactive substances binding to antibodies against AFB have also been found in urine samples from France [5] and Denmark [13]. Very sensitive and fast techniques are now available for the determination of nanogram or picogram levels of aflatoxins in human samples [14]. However, the genotoxic significance of the low levels of aflatoxins generally found in agricultural commodities or in human tissues and excreta may be questioned. Evidence of genotoxic effects from low levels of AFB is needed in order to assess the possible role of AFB in the etiology of human cancer.

There is strong evidence that the ultimate carcinogenic metabolite of AFB is the 8,9-oxide. The most abundant adduct formed between AFB and deoxyribonucleic acid (DNA) has a bond from the 7-position in guanine to the 9-position in 8-hydroxy-aflatoxin B_1 [15,16]. Due to instability of the resulting N-substituted aminoimidazole, an aflatoxin B_1 guanine adduct, $\overline{8}$,9-dihydro-8-hydroxy-9-(7'-guanyl)-aflatoxin B_1 (AFB-Gua), is released [17]. The release of AFB-Gua leads to an apurinic site in the DNA strand.

In rats, approximately 1% of an AFB dose was excreted within 24 hours as AFB-Gua [18]. Excretion of AFB-Gua from AFB-exposed humans living in an area with high AFB exposure has recently been detected by the use of HPLC and synchronous fluorescence spectroscopy [19].

The purpose of this chapter is to discuss the methods applied to evaluate genotoxic exposures to AFB and to give a report of the ongoing study on urinary excretion of AFB-Gua in different areas of Kenya.

MATERIALS AND METHODS

Chemicals

AFB-Gua was prepared as described by Martin and Garner [20], using AFB and calf thymus DNA (Sigma Chemical Company, St. Louis, MO) and ^{3}H-AFB (1.3 Ci/mmol) (Moravek Biochemicals, Brea, CA). Briefly, 1.25 mg AFB and 500 uCi ^{3}H-AFB in 0.5 ml dimethylsulfoxide were mixed with 1.6 mg calf thymus DNA in 8 ml phosphate buffer (20 mM sodium phosphate, pH 6.0). Three mg 3-chloroperoxybenzoic ac$\overline{i}$d (Merck, Darmstadt, W. Germany) was added in 8 ml dichloromethane and the reaction mixture shaken vigorously for five hours in the dark at room temperature. Unreacted AFB was removed from the water phase by three extractions with chloroform. LiCl was added to a final concentration of 1 $\underline{M}$, and the DNA was precipitated by adding 3 volumes of 96% ethanol. Excess unreacted ^{3}H-AFB was washed

off with ethanol, and the DNA redissolved in 15 nM sodium cit-
rate, pH 5.0. DNA was hydrolyzed in 0.1 N HCl at 100° C for 20
min, in order to liberate bases and base adducts. AFB-Gua was
subsequently isolated from the hydrolyzate as described in the
HPLC section. All other chemicals used were commercial grade.

Urine samples

Urine samples (25 ml or more) were collected at the out-
patient clinics at Murang'a, Makueni, and Machacos district
hospitals, and at the Kenyatta National Hospital in Nairobi,
using sterile disposable containers. Only subjects seeking
medical care for malaise with no history of liver injury were
included in the study. The samples were kept at 4° C until
further processing.

Isolation of AFB-Gua

AFB-Gua was isolated as previously described [19,21].
Briefly, the urine was adjusted to pH 5, 7% methanol was added,
and samples were centrifuged to precipitate particulates. C-18
Sep-Pak cartridges (Waters Assocs., Milford, MA) were activated
with methanol and equilibrated with 7% methanol before adsorp-
tion of urine samples. The cartridges were kept at -70° C and
shipped to either the Laboratory of Human Carcinogenesis, National Cancer Institute, Bethesda, MD, or the Laboratory of Environmental Carcinogenesis, Copenhagen, Denmark, for further
processing. The cartridges were washed with 5 ml of 10% ethanol and 5 ml of 7% acetonitrile, followed by elution with 10 ml
80% methanol. The eluate was concentrated by evaporation to
0.5 ml, and analyzed for AFB-Gua by a two-step HPLC procedure.

HPLC

The HPLC system used was composed of two model 202 pumps,
an 802 manometric module, and an 811 dynamic mixer, all from
Gilson (Villiers Le Bel, France) connected with a Gilson 502
contact module to an Apple IIe computer (Apple Inc., Cupertino,
CA). Samples were injected manually through a Rheodyne 7125
and eluates were monitored for absorption at 365 nm with a
Waters (Milford, MA) model 440 absorbance detector connected to
an LKB (Bromma, Sweden) recorder, and collected with a Gilson
microdol TDC 80 fraction collector. In the first step, isocratic elution using a C18 Bondapak™ column (Waters, Milford, MA) and 18% ethanol, 10 mM ammonium formate, pH 5.1 at a

flow rate of 1 ml/min was used. Authentic AFB-Gua eluted at 22 min and fractions eluting at 20 to 25 min were collected. Several other substances from the urine samples also eluted during this time interval. In the second step, isocratic elution of an Ultrasil™-Si (Altex, Berkeley, CA) column with 4.5% acetonitrile at 1 ml/min yielded the AFB-Gua at 5.5 min (see Figure 1). Daily ^{3}H-AFB-Gua samples, obtained by hydrolysis of AFB-DNA adducts, were run in parallel to assure stability of procedures.

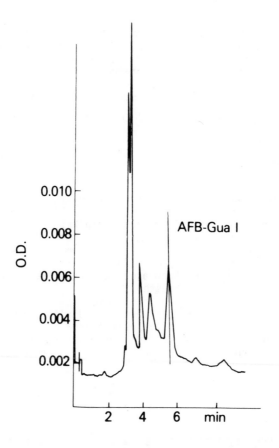

Figure 1. HPLC profile (UV, 365 nm) of urine sample positive for AFB-Gua on Ultrasil-Si column, eluted with 4.5% acetonitrile. Retention time of chemically synthesized AFB-Gua is shown by vertical line. Reprinted from "Detection of Putative Adduct with Fluorescence Characteristics Identical to 2,3-Dihydro-2-(7-Guanyl)-3-Hydroxyaflatoxin B," by Herman Autrup et al., in <u>Carcinogenesis,</u> 4:1193-1195 (1983). Copyright 1983 by IRL Press.

Synchronous Fluorescence Spectroscopy

Further identification of AFB-Gua isolated from the urine samples was achieved by synchronous fluorescence spectroscopy using a model MPF 44B spectrophotometer (Perkin-Elmer, Norwalk, CT) with synchronous luminescence and photon counting. Scanning with a fixed wavelength difference of 34 nm and a 5 nm bandwidth from 250 nm to 600 nm yielded a characteristic spectrum with a single peak at 415 nm [22]. A sample was considered positive for AFB-Gua if 365 nm absorption peaks were obtained in both HPLC systems and the characteristic synchronous fluorescence spectrum was obtained with the fraction collected from the second HPLC run.

RESULTS

At present, a total of 355 urine samples has been collected and analyzed. The age and sex distribution of the total and the positive cases are shown in Table 1. All age groups of both sexes are well represented in the study. There is a trend towards a higher number of positives among males; however, the difference is not significant.

Table 1. Age and Sex Distribution.

Age Group	Males		Females	
	Number of Cases	Percent of Cases Positive for AFB-Gua	Number of Cases	Percent of Cases Positive for AFB-Gua
11-20	38	7.9	37	10.8
21-30	36	8.3	73	6.8
31-40	24	12.5	36	8.3
41-50	12	0.0	25	0.0
51-60	17	11.8	19	0.0
61	19	15.8	19	0.0
Total	146	9.0	209	5.5

The total number of positive cases was 24 (6.7%). Several urine samples collected from volunteers at the analytical laboratories were spiked with ^{3}H-AFB-Gua (2-5 fmol/25 ml) and the recovery throughout the procedure from concentration on Sep-Pak cartridge to collection of fractions from the second HPLC run was determined. Recovery of AFB-Gua varied, but it was always better than 75%. Solutions of AFB-Gua used as standards were stable for months, when kept at -20° C. Semi-quantitative determination of the amount of AFB-Gua in the positive samples by integration of the area under the synchronous fluorescence emission peak compared to standards indicate a level of 0.3-3 pmol AFB-Gua in a 25 ml sample of urine. In Figure 2, synchronous fluorescence spectra of a standard compared to a sample considered positive are shown. The levels found would correspond to a daily excretion among positive cases of 4-40 ng AFB-Gua, assuming 75% recovery throughout the isolation procedure and a daily urine volume of one liter. Lower detection limits for AFB-Gua would then be approximately 12 fmol/ml.

The number of samples collected from the different sampling locations is shown in Table 2, together with seasonal variations in the number of positives found. All positive samples

Table 2. Analysis of Urine Samples.

Collection Site	Number of Samples Analyzed	Cases Positive[a]
Kenyatta National Hospital, Nairobi		
January-March	128	11 (8.6%)
Murang'a District Hospital		
January-March	61	7 (11.5%)
July-September	32	1 (3.1%)
Machacos District Hospital		
April-June	15	1 (7%)
Makueni District Hospital		
April-June	119	4 (3%)
Total	355	24 (6.7%)

[a]Limit of detectability was 0.3 pmol.

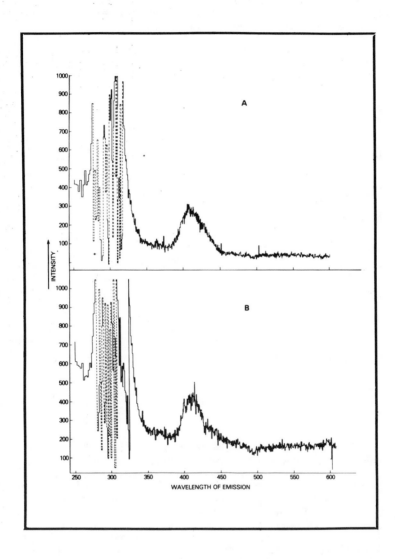

Figure 2.   Synchronous fluorescence emission spectrum for chemically synthesized AFB-Gua (A) and positive test sample (B - same sample as in Figure 1). Reprinted from "Detection of Putative Adduct with Fluorescence Characteristics Identical to 2,3-Dihydro-2-(7-Guanyl)-3-Hydroxyaflatoxin B," by Herman Autrup et al., in Carcinogenesis, 4:1193-1195 (1983). Copyright 1983 by IRL Press.

collected at the Kenyatta National Hospital were found to derive from individuals residing in the Kiambu, Machacos, Meru/
Embu and Murang'a districts or from the more rural suburban
areas of Nairobi as shown in Table 3.

Table 3.   Analysis of AFB-Gua in Urine Samples Collected at
Kenyatta National Hospital.

District	Number of Samples Analyzed	AFB-Gua Positive Cases[a]
Kiambu	39	3
Machacos	11	0
Meru/Embu	9	3
Murang'a	25	5
Others, including Nairobi	44	0
Total	128	11

[a]Limit of detectability was 0.3 pmol.

DISCUSSION

The relationships between human exposures to carcinogens
and subsequent cancer rates are hard to establish, partly because large inter-individual differences in activation, DNA
binding, and DNA repair of carcinogens exist in human populations. Indication that such differences in genotoxic susceptibility to AFB between individuals exist comes from in vitro
studies with human tissue cultures. Following in vitro incuba-
tion of human bronchial or colonic explants with AFB, a 100-
fold variation among individuals was found in the formation of
adducts between AFB and DNA [16].
Also, the carcinogenic effect of human aflatoxin exposures
cannot be extrapolated using dose-response relationships obtained from animal experimentation. Correlation between doses
of AFB and cancer rates in different animal species is generally poor. However, when carcinogen-DNA binding in target organs
is measured, good correlation with cancer rates has been observed for several carcinogens, including AFB [23]. Therefore,
determination of the binding of AFB to DNA in humans should

give a better basis for extrapolation of cancer risks from animal data. For most carcinogens, including AFB, this requires the development and application of highly sensitive analytical methods.

An AFB adduct to the 7-position of guanine in DNA leads to destabilization of the imidazole ring structure of guanine, due to the delocalization of a positive charge. This destabilized structure may react in any of three ways. One is by release of the 8,9-dihydrodiol derivative of AFB (AFBdiol), leaving an intact guanine residue in DNA. This pathway is, however, a minor one, as only a little AFBdiol was found to be released from AFB-adducted DNA [24]. A second way of decomposition is by the release of the complete AFB-Gua moiety, leaving an apurinic site in the DNA strand [25], which is a possible mutagenic lesion [26]. In eucaryotes, apurinic sites have been shown to lead to TA-GC transversions [27]. This particular change in base composition has been found in several activated oncogenes isolated from human tumor cell lines [28,29]. A third way of hydrolysis leads to opening of the positively charged imidazole ring structure, with the formation of an adduct identified as 8,9-dihydro-9-(N^5-formyl-2',5',6'-triamino-4'-oxo-N^5-pyrimidyl)-9-hydroxy-aflatoxin B$_1$ [30]. No repair of this lesion was found after 72 hours in rat liver [31].

The half-life of AFB-Gua adducts was found to be around 10 hours in DNA prepared from livers of AFB-dosed rats [31], in DNA from AFB-exposed human fibroblast cultures [32], and in synthetically prepared AFB-DNA at physiological pH [33]. Hydrolysis of adducts was found to proceed mainly through spontaneous release of AFB-Gua. The similar half-lifes for AFB-DNA found in these different systems, including adducts produced chemically, indicate that active DNA repair plays a minor role. Furthermore, the three hydrolytic pathways all seem to proceed spontaneously under physiological conditions. After a single dose of AFB in the rat, 10-20% of initially bound AFB remains as persistent ring-opened adducts in liver DNA. The level of persistent adducts increases with every subsequent dose given [31].

It can be argued that part of the AFB-Gua found in urine samples may be released from ribonucleic acid (RNA) adducts or from DNA adducts in other organs than the liver. In the rat, there was a good correlation between initial AFB-DNA adduct levels in the liver at different dose levels and excretion of AFB-Gua in urine within 48 hours [17,19]. Approximately one third of the initially bound AFB was found in rat urine within this time interval. Thus, in the rat there is good correlation between initial DNA binding, persistent adducts in DNA, and spontaneous release of AFB-Gua. Furthermore, the release of AFB-Gua is mainly by spontaneous mechanisms, making urinary adduct levels an ideal indicator for determination of in vivo DNA binding rates. Assuming similar rates of release and excretion of AFB-Gua from liver AFB-DNA adducts in rat and man, i.e., 30-40% initial AFB-DNA adducts excreted within 48 hours, and taking into consideration differences in organ weights, human adduct levels can be estimated. From the positive cases among our present results, a daily introduction of 0.02-0.4 AFB-DNA adducts/10^7 nucleotides in the liver can be antici-pated. In this calculation, human liver weights of 1.5 kg, similar DNA contents per g of liver in rat and man, and daily human urine excretion of 0.5-1 liter were assumed.

The level of AFB-DNA adducts per unit dose varies widely between animal species, but there is good agreement between initial binding rates and cancer incidences in similarly dosed animals [2,19]. If a similar relation holds for humans, the present study clearly indicates an important role for AFB in the etiology of human liver cancer.

Recovery of ^{3}H-AFB-Gua at low concentrations (2-5 pmol/25 ml) in spiked urine samples was quite variable, but never was below 75%. Loss was mainly due to the Sep-Pak preconcentration procedure. A recovery of 69% AFB-Gua from 60 ml samples con-taining 25 ng of the compound has been reported [18], whereas almost quantitative recoveries were reported at higher concen-trations. None of our samples were spiked with AFB-Gua in Kenya, but the stability of AFB-Gua makes it quite unlikely that recoveries in the study should differ much from the num-bers found experimentally.

The present study confirms earlier studies on the geo-graphical distribution of AFB-contaminated food and AFB expo-sures in Kenya [4]. Furthermore, the findings indicate that human consumption of AFB-contaminated food leads to activation of ingested AFB and subsequent binding of the ultimate carcin-ogen to nucleic acids. The highest number of positives were found in the Murang'a district in the period January to March, when the maize and beans that constitute the major food items have been stored for some time. Relatively low numbers of positives were found in the period April to June, when most of the food is bought from government storage depots, and in the post harvest season, July to September.

In the city of Nairobi, most food items are bought from stores with good storage facilities. Consquently, minimal AFB exposure of the population was expected, and the urine samples collected at the Kenyatta National Hospital in Nairobi were meant to serve as negative controls for the highly exposed rural population. The relatively large numbers of positives found among patients from the Kenyatta National Hospital may be explained by the fact that patients from rural areas around Nairobi City were included.

Only a few studies have aimed at relating human exposure to binding levels of carcinogens in DNA. A main problem is the lack of good indicators for the determination of the dose to human target organ DNA. Measurements of DNA adducts of benzo(a)pyrene (BP-DNA) in human lymphocytes in roofers and foundry workers by enzyme-linked immunosorbent assay (ELISA) or by ultrasensitive enzyme radioimmunoassay (USERIA) techniques gave evidence of the existence of adducts in several of the subjects screened due to occupational exposures [34]. In non-occupationally exposed controls, two positive cases could be related to smoking [34]. A non-significant increase in BP-DNA adducts, measured by ELISA assays in placentas of smokers as compared with nonsmokers, has been reported recently [35]. Very interestingly, an unknown adduct strongly related to smoking was reported in the same study by the use of the 32P postlabeling assay. The use of synchronous fluorescence spectroscopy to assess BP-DNA has also been described [36]. This method was compared with USERIA in the assessment of BP-DNA in lymphocytes from coke oven workers [37], and a reasonable number of positive or negative cases were identified by both methods. Furthermore, several workers were found to produce antibodies to BP-DNA, indicating new possibilities in the assessment of exposures [37].

There is an urgent demand for the development of sensitive and practical techniques to assess genotoxic exposures to specific carcinogens in man. The present study is one of the first of its kind and clearly indicates that AFB reacts with nucleic acids following human ingestion, and that AFB may be of importance as a possible risk factor for human liver cancer in countries where AFB-contaminated food is used for human consumption.

ACKNOWLEDGMENTS

The authors would like to thank Dr. Kirsi Vähäkangas for valuable help with synchronous fluorescence measurements. The work was supported in part by a grant from the Neye Foundation to Herman Autrup and by a core grant from the Danish Cancer Society to the Fibiger Institute. The work in this chapter was not funded by EPA and no official endorsement should be inferred.

REFERENCES

1. Stoloff, L. "Aflatoxins - An Overview," in Mycotoxins in Human and Animal Health, J. V. Rodricks, C. W. Hesseltine and M. A. Mehlman, Eds. (Park Forest South: Pathotox Publishers, Inc., 1977), p. 7.

2. Busby, Jr., W. F., and G. N. Wogan. "Aflatoxins," in Chemical Carcinogens, 2nd ed., C. E. Searle, Ed. (Washington, D.C.: American Chemical Society, 1984), p. 945.

3. "Aflatoxins," in IARC Monographs on the Evaluation of Carcinogenic Risk of Chemicals to Man. Some Naturally Occurring Substances (Lyon: International Agency for Research on Cancer, 1976), p. 51.

4. Peers, F. G. "Dietary Aflatoxins and Liver Cancer - A Population Based Study in Kenya," Br. J. Cancer 27:473-483 (1973).

5. Martin, C. N., R. C. Garner, F. Tursi, J. V. Garner, H. C. Whittle, R. W. Ryder, P. Sizaret, and R. Montesano. "An Enzyme-Linked Immunosorbent Procedure for Assaying Aflatoxin B₁," in Monitoring Human Exposure to Carcinogenic and Mutagenic Agents, Berlin, A., M. Draper, K. Hemminki, and H. Vainio, Eds. (Lyon: International Agency for Research on Cancer, 1984), p. 313.

6. Onyemelukwe, C. G., and G. Ogbadu. "Aflatoxin Levels in Sera of Healthy First Time Blood Donors: Preliminary Report," Trans. Royal Soc. Trop. Med. Hyg. 75:780-782 (1981).

7. Onyemelukwe, C. G., G. Ogbadu, and A. Salifu. "Aflatoxins B₁, B₂, G₁, G₂ in Primary Liver Cell Carcinoma," Toxicol. Lett. 10:309-312 (1982).

8. Hendrickse, R. G., J. B. S. Coulter, S. M. Lamplugh, S. B. J. Macfarlane, T. E. Williams, M. I. A. Omer, and G. I. Suliman. "Aflatoxins and Kwashiorkor: A Study in Sudanese Children," Brit. Med. J. 285:843-846 (1982).

9. Shank, R. C., C. H. Bourgeois, N. Keschamras, and P. Chandavimol. "Aflatoxins in Autopsy Specimens from Thai Children with an Acute Disease of Unknown Aetiology," Fd. Cosmet. Toxicol. 9:501-507 (1971).

10. Siray, M. Y., A. W. Hayes, P. D. Unger, G. R. Hogan, N. J. Ryan, and B. B. Wray. "Analysis of Aflatoxin B₁ in Human Tissues with High-Pressure Liquid Chromatography," Toxicol. Appl. Pharmacol. 58:422-430 (1981).

11. Tsuboi, S., T. Nakagawa, M. Tomita, T. Seo, H. Ono, K. Kawamura, and N. Iwamura. "Detection of Aflatoxin B₁ in Serum Samples of Male Japanese Subjects by Radioimmunoassay and High-Performance Liquid Chromatography," Cancer Res. 44:1231-1234 (1984).

12. Yang, G., S. Nesheim, J. Benavides, I. Ueno, A. D. Campbell, and A. Pohland. "Radioimmunoassay Detection of Aflatoxin B₁ in Monkey and Human Urine," Zbl. Bakt. Suppl. 8:329-335 (1980).

13. Dragsted, L. O., A. I. Larsen, I. Bull, and H. Autrup. "Excretion of Aflatoxins in Dockers with Occupational Exposure," Ugeskr. Læger. 147:4148-4150 (1985).

14. Garner, C., R. Ryder, and R. Montesano. "Monitoring of Aflatoxins in Human Body Fluids and Application to Field Studies," Cancer Res. 45:922-928 (1985).

15. Essigman, J. M., R. G. Croy, A. M. Nadzan, W. F. Busby, V. N. Reinhold, G. Buchi, and G. N. Wogan. "Structural Identification of the Major DNA Adduct Formed by Aflatoxin B1 in vitro," Proc. Natl. Acad. Sci. U.S.A. 74:1870-1874 (1977).

16. Autrup, H., J. M. Essigman, R. G. Croy, B. F. Trump, G. N. Wogan, and C. C. Harris. "Metabolism of Aflatoxin B1 and Identification of the Major Aflatoxin B1-DNA Adducts Formed in Cultured Human Bronchus and Colon," Cancer Res. 39:694-698 (1979).

17. Essigman, J. M., R. G. Croy, R. A. Bennett, and G. N. Wogan. "Metabolic Activation of Aflatoxin B1: Patterns of DNA Adduct Formation, Removal, and Excretion in Relation to Carcinogenesis," Drug Metab. Reviews 13:581-602 (1982).

18. Bennett, R. A., J. M. Essigman, and G. N. Wogan. "Excretion of an Aflatoxin-Guanine Adduct in the Urine of Aflatoxin B1 Treated Rats," Cancer Res. 41:650-654 (1981).

19. Autrup, H., K. A. Bradley, A. K. M. Shamsuddin, J. Wakhisi, and A. Wasunna. "Detection of Putative Adduct with Fluorescence Characteristics Identical to 2,3-Dihydro-2-(7'-guanyl)-3-hydroxyaflatoxin B1 in Human Urine Collected in Murang'a District, Kenya," Carcinogenesis 4:1193-1195 (1983).

20. Martin, C. N. and R. C. Garner. "Aflatoxin B-oxide Generated by Chemical or Enzymatic Oxidation of Aflatoxin B1 Causes Guanine Substitution in Nucleic Acids," Nature (London) 267:863-865 (1977).

21. Autrup, H., J. Wakhisi, K. Vähäkangas, A. Wasunna, and C. C. Harris. "Detection of 8,9-Dihydro-(7'-guanyl)-9-hydroxy-aflatoxin B1 in Human Urine," Environ. Hlth. Perspect. 62:105-108 (1985).

22. Vo-Dinh, T. "Multicomponent Analysis by Synchronous Luminescence Spectrometry," Anal. Chem. 50:396-401 (1978).

23. Neumann, H.-G. "Dosimetry and Dose-Response Relationships," in Monitoring Human Exposure to Carcinogenic and Mutagenic Agents, A. Berlin, M. Draper, K. Hemminki, and H. Vainio, Eds. (Lyon: International Agency for Research on Cancer, 1984) p. 115-126.

24. Lin, J.-K., J. A. Miller, and E. C. Miller. "2,3-Dihydro-2-(guan-7-yl)-3-hydroxy-aflatoxin B_1, a Major Acid Hydrolysis Product of Aflatoxin B_1-DNA or -Ribosomal RNA Adducts Formed in Hepatic Microsome- mediated Reactions and in Rat Liver in Vivo," Cancer Res. 37:4430-4438 (1977).

25. Croy, R. G., J. M. Essigman, V. N. Reinhold, and G. N. Wogan. "Identification of the Principal Aflatoxin B_1-DNA Adduct Formed in Vivo in Rat Liver," Proc. Natl. Acad. Sci. U.S.A. 75:1745-1749 (1978).

26. Stark, A. A., J. M. Essigman, A. L. Demain, T. R. Skopek, and G. N. Wogan. "Aflatoxin B_1 mutagenesis, DNA-binding, and Adduct Formation in Salmonella typhimurium," Proc. Natl. Acad. Sci. U.S.A. 76:1343-1347 (1979).

27. Forster, P. L., E. Eisenstadt, and J. H. Miller. "Substitution Mutations Induced by Metabolically Activated Aflatoxin B_1," Proc. Natl. Acad. Sci. U.S.A. 80:2695-2698 (1983).

28. Reddy, E. P., R. K. Reynolds, E. Santos, and M. Barbacid. "A Point Mutation is Responsible for the Acquisition of Transforming Properties by the T24 Human Bladder Carcinoma Oncogene," Nature 300:149-152 (1982).

29. Capon, D. J., P. H. Seeburg, J. P. McGrath, J. S. Hayflick, U. Edman, A. D. Levinson, and D. V. Goeddel. "Activation of Ki-ras 2 Gene in Human Colon and Lung Carcinomas by Two Different Point Mutations," Nature 304:507-513 (1983).

30. Hertzog, P. J., J. R. Lindsay Smith, and R. C. Garner. "A High Pressure Liquid Chromatography Study on the Removal of DNA-bound Aflatoxin B_1 in Rat Liver and In Vitro," Carcinogenesis 1:787-793 (1980).

31. Croy, R. G., and G. N. Wogan. "Temporal Patterns of Covalent DNA Adducts in Rat Liver after Single and Multiple Doses of Aflatoxin B_1," Cancer Res. 41:197-203 (1981).

32. Leadon, S. A., R. M. Tyrell, and P. A. Cerutti. "Excision Repair of Aflatoxin B_1-Adducts in Human Fibroblasts," Cancer Res. 41:5125-5129 (1981).

33. Groopman, J. D., R. G. Croy, and G. N. Wogan. "In Vitro Reactions of Aflatoxin B_1-Adducted DNA," Proc. Natl. Acad. Sci. U.S.A. 78:5445-5449 (1981).

34. Shamsuddin, A. K. M., N. T. Sinopoli, K. Hemminki, R. R. Boesch, and C. C. Harris. "Detection of Benzo(a)pyrene-DNA Adducts in Human White Blood Cells," Cancer Res. 45:66-68 (1985).

35. Everson, R. B., E. Randerath, R. M. Santella, R. C. Cefalo, T. A. Avitts, and K. Randerath. "Detection of Smoking-Relating Covalent DNA Adducts in Human Placenta," Science 231:54-57 (1986).

36. Vähäkangas, K., A. Haugen, and C. C. Harris. "An Applied Synchronous Fluorescence Spectrophotometric Assay to Study Benzo(a)pyrene-diolepoxide-DNA Adducts," Carcinogenesis 6:1109-1116 (1985).

37. Harris, C. C., K. Vähäkangas, M. J. Newman, G. E. Trivers, A. Shamsuddin, N. Sinopoli, D. L. Mann, and W. E. Wright. Personal communication (1986).

CHAPTER 2

ASSESSMENT OF HUMAN EXPOSURE TO CHEMICALS
THROUGH BIOLOGICAL MONITORING

Alfred M. Bernard and Robert R. Lauwerys

INTRODUCTION

Traditionally, the assessment of human exposure to chemicals mainly relies on environmental monitoring. The latter evaluates the potential exposure, i.e., the amount of chemicals likely to reach the respiratory tract or to be absorbed by the organism depending on several factors such as the physico-chemical properties of the substance, the hygiene habits of the worker, or some biological factors (e.g., age, sex, ventilatory parameters). Studies on the fate of chemicals in the human organism and on their biological effects have led to various methods for exposure monitoring grouped under the name biological monitoring of exposure. The main advantage of this approach is to provide, for chemicals acting systematically, a better assessment of health risk than the environmental measurements. The objective of this chapter is to review the available biological methods and their main applications in the field of occupational and environmental medicine.

DEFINITION AND ROLE OF BIOLOGICAL MONITORING OF EXPOSURE

The objective of biological monitoring (BM) of exposure is basically the same as that of ambient monitoring, i.e., to prevent excessive exposures to chemicals which may cause acute or chronic adverse health effects. In both approaches, the health risk is assessed by comparing the value of the measured parameter with its currently estimated maximum permissible value in the analyzed medium (threshold limit value [TLV] or biological limit value [BLV]). BM of exposure, like ambient monitoring, is essentially a preventive activity and in this respect, it must be clearly distinguished from BM of effects (also called health surveillance), which, by means of sensitive biological markers, aims at detecting - and not preventing - the early signs of toxicity [1,2].

But while ambient monitoring attempts to estimate the external exposure to a chemical, BM directly assesses the amount of chemical effectively absorbed by the organism, i.e., the internal dose. Depending on the characteristics of the selected biological parameter (particularly its biological half-life) and the conditions under which it is measured, the term internal dose may have different meanings, such as the total amount or a fraction (e.g., biologically active dose) of chemical recently absorbed (recent exposure), the amount stored in one or several body compartments (total integrated exposure or specific organ dose), or the amount bound to the target sites (target dose). It is thus evident that contrary to environmental monitoring, which only assesses the amount of chemical reaching the exposed organism by one or several routes at the time of sampling or during a certain time interval (continuous monitoring), BM methods may estimate fractions of the internal dose with various biological significances.

BM of exposure is usually reserved for chemicals that penetrate into the organism and exert systemic effects. Very few biological tests have been proposed for the identification or the monitoring of chemicals entering the interface between the environment and the organism (skin, gastrointestinal mucosa, respiratory tract mucosa). The analysis of nickel in nasal mucosa and the counting of asbestos bodies in sputum could be considered as examples of such tests. For systemically active chemicals, BM of exposure represents the most effective approach for assessing the potential health risk, since a biological index of internal dose is necessarily more closely related to a systemic effect than any environmental measurement.

BM of exposure integrates the chemical absorption by all routes (pulmonary, oral, cutaneous) and from all possible sources (occupational, environmental, dietary, etc.). This property is particularly useful when assessing the overall exposure to widely dispersed pollutants. Even for elements present in the environment under different chemical forms with different toxicities (e.g., inorganic arsenic in water or in industrial settings and organic arsenic in marine organisms), it may still be possible to correctly estimate the health risk by speciation of the element in the analyzed biological medium. BM of exposure takes into account the various individual factors which influence the uptake or the absorption of the chemical (e.g., sex, age, physical activity, hygiene, nutritional status, etc.). In general, the meaningful application of a biological test for determining the internal dose of a chemical requires the collection of relevant information on its metabolism (absorption, distribution, excretion), its toxicity, and on the relationships between internal dose, external exposure, and adverse effects. The knowledge of the latter permits one to estimate directly or indirectly (from the TLV) the maximum permissible internal dose (BLV) [1,2]. Unfortunately, for many industrial chemicals, one or all of the preceding conditions are not fulfilled, which limits the possibilities of BM. As mentioned above, BM is usually not applicable to substances acting locally. This approach is also not useful for detecting peak exposures to rapidly acting substances. The detection of excessive exposure to these chemicals should mainly rely on the continuous monitoring of the pollutant concentration in the environment. Finally, some BM tests may be sensitive to various confounding factors of endogenous or exogenous origin, which may lead to an erroneous interpretation of the results.

BIOLOGICAL TESTS OF EXPOSURE

Tests Measuring the Chemical or its Metabolites in Biological Media

The majority of biological tests currently available for monitoring exposure to chemicals rely on the determination of the chemical or its metabolites in biological media. Urine, blood, and alveolar air are the most commonly used media. The analysis of other biological materials such as milk, fat, saliva, hair, nails, teeth, and placenta is less frequently performed.

As a general rule, urine is used for inorganic chemicals and for organic substances which are rapidly biotransformed to more hydrosoluble compounds, blood is used for most inorganic chemicals and organic substances poorly biotransformed, and alveolar air analysis is reserved for volatile compounds (e.g., solvents). The measured parameter and the time of sampling must be selected by considering the physico-chemical properties of the substance, the exposure conditions, several toxicokinetic parameters (distribution, biotransformation, elimination), the sensitivity of the analytical methods, and also the type of information required (e.g., recent exposure, body burden, organ dose, target dose). In the case of cadmium, for instance, the concentration of the metal in whole blood may be mainly influenced either by the cadmium body burden (e.g., in workers removed from exposure for several years), or by the last few months' exposure (e.g., workers currently exposed to levels exceeding 10 g/m^3), while its urinary excretion is a good index of the amount accumulated in the kidneys [3,4].

For chemicals which must be activated before reaching the target sites, the determination of the active metabolite or of a metabolite deriving from the activated form may be more relevant for the health risk assessment than that of the parent compound or of any other metabolite not involved in the toxic process. For example, the analysis of hexanedione, the metabolite responsible for the neurotoxicity of n-hexane, might be more useful than that of n-hexanol in urine or n-hexane in expired air for monitoring the exposure to this solvent [2].

Tests based on the determination of the chemical or its metabolites in biological media may be selective or non-selective. Selective tests are those measuring a well-defined chemical, while non-selective tests evaluate the exposure to a group of chemicals (e.g., azo derivatives in urine). BM tests may also be invasive (i.e., requiring a sample of blood or tissue), or non-invasive (i.e., tests analyzing urine, alveolar air, hair, etc.). Particularly interesting are the non-invasive methods developed recently for measuring in vivo the metal content of selected tissues. These methods, usually based on neutron activation or on x-ray fluorescence techniques, have already been successfully applied to the determination of cadmium in kidney or liver or of lead in bones [for a review see 5].

Tests Based on the Determination of a Non-adverse Biological Effect Related to the Internal Dose

A biological effect is considered as non-adverse if the functional or physical integrity of the organism is not diminished, if the ability of the organism to face an additional stress (homeostasis) is not decreased, or if these impairments are not likely to occur in the near future (delayed toxicity). The advantage of tests measuring a non-adverse biological effect is that they may provide information on the amount of chemical likely to react with the target sites. The determination of alkylated hemoglobin or of erythrocyte cholinesterase activity are tests based on this principle.

In some cases, however, the non-adverse biological effect has no more predictive value than the mere determination of the chemical itself. For instance, in the BM of exposure to cadmium, the analysis of metallothionein in urine seems to offer no other advantage over that of cadmium except of not being sensitive to the external contamination [6].

Tests Measuring the Amount of Chemical Bound to the Target Molecules

The most useful BM methods are those directly measuring the amount of active chemical bound to the target molecules (target dose). When feasible (i.e., when the target site is readily accessible), these methods may assess the health risk more accurately than any other monitoring test. The carboxyhemoglobin test, in application in industry for several decades, belongs to this category. Progress in this monitoring approach is to be expected, namely in the field of genetic toxicology, where immunoassays are currently being developed for measuring adducts between DNA and various carcinogens or mutagens.

AREAS OF APPLICATION

Routine Exposure Monitoring in Industry

The main objective of the biological methods listed above is the accurate evaluation of the internal dose of a chemical in view of assessing the potential health risk. In Europe, at least, the major burden of the biological tests currently performed by occupational health services attempts to meet this objective. Tentative biological exposure limits have been proposed by different organizations. This application, however, is still restricted to a few chemicals, because as stressed above, all the conditions required to propose meaningful biological limit values are not always fulfilled. But even when the available information is too limited to interpret the results of the biological tests in terms of exposure intensity or health risk, it may still be useful to perform them for other purposes, as listed below.

Research on Associations Between Chemical Exposure and Health Effects

A causal association between health impairment and excessive exposure to a chemical may be suggested by the finding of abnormally elevated concentrations of the chemical in the organism. For instance, the pathological role of aluminum in the osteomalacia and encephalopathy of dialysis patients was initially suggested by the finding of tremendous accumulations of aluminum in the bones and brains of these patients. The source of aluminum was clearly identified when it was found that the degree of plasma and/or tissue aluminum accumulation was positively related to the duration of hemodialysis treatment or to the aluminum concentration in the water supplier [7].

Similarly, the diagnosis of an anemia or a nephropathy caused by an occult plumbism relies mainly on the determination of the lead body burden (e.g., the ethylene diamine tetracetic acid $CaNa_2$ [EDTA]-lead mobilization test) [8]. During cross-epidemiological studies, BM data may also help in the matching of groups and in excluding the possible interference of confounding factors. We have recently examined the fertility of male workers exposed to manganese dust. BM tests applied to

blood and urine were used to ascertain that the examined workers were not simultaneously exposed to cadmium, mercury, or
lead but only to manganese. The fertility of these workers was
found to be significantly depressed during their exposure to
manganese, which strongly suggests a causal association between
excessive exposure to this metal and impaired reproductive performance [9].

However, some caution is required in the evaluation of the
causal nature of a relationship between chemical exposure and
health effects. The latter, indeed, may be the cause rather
than the consequence of an excessive internal dose of a chemical. The accumulation of aluminum in patients with renal insufficiency treated by dialysis is an example of increased uptake of a chemical caused by a previous disease state, although
the progressive accumulation of the metal may eventually lead
to the occurrence of other adverse effects. In the study of
the association between lead exposure and both renal insufficiency and hypertension, it was also considered that renal
impairment might be responsible for the elevation of blood lead
concentrations by decreasing the urinary excretion of the metal. But even in patients with severe renal failure, the renal
clearance of lead was not affected, which suggests that the increased lead body burden could be an etiologic factor rather
than a mere consequence of these diseases [10].

The association between a chemical and a health effect may
also be secondary (i.e., non-causal). A typical example of
such an association is the presence of high levels of cadmium
in tissues (e.g., lungs, liver, or kidneys) of persons deceased
of lung cancer, emphysema, or chronic bronchitis. This association, which was reported in the past as possibly causal, is
better explained by the fact that tobacco smoke may contain
high levels of cadmium, and tobacco consumption is a well known
etiologic factor in these diseases [6].

Establishment of Dose-response Relationships

Dose-response or dose-effect relationships (i.e., relationships between the frequency or the intensity of health effects
and internal exposure) may sometimes constitute an argument
supporting the existence of a causal association, despite the
fact that they may be observed in non-causal associations
(e.g., cadmium in tissues and the incidence of lung cancers in
smokers), and that in some cases the effect is not related to
the internal dose over the entire range of exposure. The
greatest interest of these relationships is the fact that they
allow the suggestion of biological limit values (BLV). Cross-
sectional studies performed among populations at risk and using
sensitive indicators of health effects represent the most prag-
matic approach to establish dose-response relationships. Such
studies have enabled us to propose BLVs for occupational expo-

sure to mercury vapor and cadmium. Prolonged exposure to cadmium results in the progressive accumulation of this metal in the organism, mainly in the liver and the kidneys. The latter is usually considered as the critical organ, i.e., the first organ to be injured. Renal dysfunction induced by cadmium can be detected at an early stage by measuring specific urinary proteins such as albumin, retinol-binding protein, and $_2$-microglobulin. The accumulation of cadmium in the kidneys can be directly assessed in vivo by neutron activation analysis or indirectly from the urinary excretion of cadmium. On the basis of the relationships between the indicators of renal impairment and the cadmium body burden established in male industrial workers, we have proposed BLV for the concentration of cadmium in urine (10 g/g creatinine), in renal cortex (216 ppm), and in liver (30 ppm) [6]. The kidneys and the central nervous system are the two critical organs during chronic exposure to inorganic mercury. In workers exposed to elemental mercury vapor, we have found that the prevalences of preclinical signs of renal dysfunction are increased mainly in subjects with a urinary excretion of mercury exceeding 50 g/g creatinine. At exposure levels below this threshold, the risk of central nervous system disturbances (e.g., tremor) is also very low [11]. Unfortunately, for many chemicals, the relationships between internal dose and adverse effects are insufficiently or even not documented. In those cases, the BLV is derived indirectly from the TLV by means of toxicokinetic data usually collected in controlled human studies [2].

Identification of Groups at Risk

BM may also be used for the identification of groups of workers exposed to certain chemicals or groups of chemicals (e.g., mutagenicity of urine), or to follow trends in exposure without necessarily assessing with precision the internal dose and the potential health risk associated with exposure. This information, however, may be useful for designing appropriate epidemiological studies. A similar approach may also be applied to the general population. The doubling of chemical consumption every seven years in industrialized societies necessarily entails a global pollution of the ecosystem with persistent hazardous chemicals such as PCB derivatives or heavy metals. In various parts of the world, projects have been undertaken for monitoring these pollutants in tissues and body fluids of populations suspected of being at risk. For instance, a collaborative project was recently carried out by United Nations Environment Program/World Health Organization (UNEP/WHO) to assess human exposure to cadmium and lead in different areas of the world. In the case of cadmium, the results show that the mean concentration of this metal in the renal cortex (i.e., the target organ) in the age group of 40 to 59 years varies between 20 and 30 ppm in the United States, Sweden, China, India, and Israel, but reaches values up to 38 ppm in Belgium and 65 ppm in Japan [12]. In Belgium, cadmium pollution is mainly localized in areas (e.g., the Liège area) where non-ferrous smelters have been in activity for many years. To determine whether this environmental pollution by cadmium may have led to a higher uptake of cadmium by the inhabitants, we have compared the cadmium level in the blood and urine of aged women who have spent the major part of their lives in the Liège area with that of a control group of women matched for age and socio-economic status and selected in an industrial area not polluted by cadmium. The urinary excretion of cadmium was found to be, on the average, twice as high in the Liège area than in the control area. Since the cadmium level in urine mainly reflects the body burden of the metal, these results indicate that on the average, elderly women from the Liège area have accumulated more cadmium in their organism than did women from the control area. The concentration of cadmium in blood was also higher in the Liège area than in the control area, which is in agreement with the current environmental pollution by cadmium [13]. These results were confirmed by a recent autopsy study in which 251 liver and 443 kidney cortex samples from the Liège area or from the remainder of the country were analyzed for their cadmium content [14]. In all age groups, the persons who had lived in the contaminated area had stored more cadmium in their livers and renal cortexes than did residents from other areas of Belgium. The same trend was found in males and females, which strengthens the hypothesis of an environmental factor.

Toxicokinetic Studies on Human Subjects

As indicated above, the knowledge of the metabolic fate of chemicals is a prerequisite for developing biological tests of exposure. Such information is usually collected on volunteer subjects in industry or under experimental exposure conditions. When the results suggest that the tests are potentially useful, additional kinetic studies may be relevant to identify possible confounding factors. Such studies have shown that ethanol can competitively inhibit the enzymatic oxidation of substances such as styrene [15] or toluene [16]. Diseases may also be a source of confounding in BM. Studies among patients with liver diseases have shown that the proportions of monomethylarsenic and dimethylarsenic acid excreted in urine following exposure to inorganic arsenic are closely related to the functional integrity of the liver [17]. Toxicokinetic studies have also shown that physical activity, body fat, site of skin contact, and drug consumption may also act as confounding factors in some BM tests [18-20].

Assessment of the Efficiency of Protective Measures and
Identification of the Main Route of Absorption

Because of its capability to evaluate absorption of chemicals by all routes, BM is particularly adapted for evaluating
the efficiency of individual protective devices such as gloves,
masks, or barrier creams. We have tested on volunteers the effect of two barrier creams containing glycerol or silicone on
the percutaneous absorption of m-xylene [21]. The absorption
of the solvent was evaluated by measuring the amount of m-
xylene eliminated in exhaled air and the 24 h urinary excretion
of methylhypuric acid. Although these creams had been validated in vitro by the manufacturers; in volunteers, they sur-
prisingly had no significant effect on the skin absorption of
m-xylene [21]. In a similar study conducted among workers exposed to dimethylformamide (DMF) in an acrylic fiber factory,
we compared the efficiency of gloves with that of a glycerol-
based barrier for preventing skin absorption of DMF [22]. The
patterns of N-methylformamide (NMF, the main metabolite of DMF)
excretion in urine observed with the different protective devices clearly showed that the use of impermeable gloves was a
more effective way for avoiding cutaneous absorption of DMF
than the use of the barrier cream. Furthermore, the comparison
of NMF excretion in urine when the workers were or were not
wearing a respiratory protective device enabled us to conclude
that the lungs did not represent an important route of entry.
However, the removal of gloves led to a marked increase of the
urinary excretion of NMF. This observation demonstrated that
the skin was the main route of exposure to DMF.

CONCLUSION

For some chemicals and under some exposure conditions, BM
offers the potential of a more accurate and reliable assessment
of uptake than ambient monitoring. For other chemicals (e.g.,
locally acting substances) or other exposure circumstances
(e.g., peak exposure), environmental monitoring may be the
method of choice for preventing health risks. It is, however,
likely that in many situations the information provided by both
monitoring approaches is complementary. However, the potential
of BM is far from being completely realized, and it can be expected that in the future this approach will further develop in
both quantitative and qualitative terms. The steady improvement of the sensitivity and specificity of analytical methods
broadens the spectrum of chemicals which can be analyzed in
biological media. Increasing automation, by reducing the duration and cost of chemical determinations, makes them more suitable to routine application. Analytical advances also improve
the quality of information which can be obtained from BM
tests. The development of methods measuring specific forms of

a chemical (analytical speciation) or evaluating the amount of
a chemical stored in the target organs or bound to the target
molecules will increase our capability to assess the toxicolog-
ically relevant internal dose and hence the health risk.   The
steady progress in the understanding of the metabolic fate and
of the mode of action of occupational or environmental pollutants may also suggest new biological indicators potentially
applicable for BM. But these promising perspectives should not
let us forget that BM of exposure uses man as an integrator of
exposure. The ethical aspects must receive a great deal of attention, and in particular, BM must always be applied under
conditions which respect some basic rights of the examined subject, such as the right to the confidentiality of the results
and the right to be informed of the risks, benefits, and results of the test.

DISCLAIMER

The work described in this chapter was not funded by EPA
and no official endorsement should be inferred.

REFERENCES

1. Lauwerys, R. Industrial Chemical Exposure: Guidelines
    for Biological Monitoring (Davis, California: Biomedical
    Publications, 1983).

2. Bernard, A., and R. Lauwerys. "General principles of bio-
    logical monitoring of exposure to organic chemicals," in
    Biological Monitoring of Exposure to Chemicals, Vol. 1,
    Organic Compounds. M. H. Ho and H. K. Dillon, Eds. (New
    York: John Wiley and Sons, 1986, in press).

3. Lauwerys, R., H. Roels, M. Regnier, J. P. Buchet, A. Ber-
    nard, and A. Goret. "Significance of Cadmium Concentra-
    tion in Blood and in Urine in Workers Exposed to Cadmium,"
    Environ. Research 20:375-391 (1979).

4. Hassler, E., B. Lind, and M. Piscator. "Cadmium in Blood
    and Urine Related to Present and Past Exposure, a Study of
    Workers in an Alkaline Battery Factory," Brit. J. Ind.
    Med. 40:420-425 (1983).

5. Lauwerys, R. "In Vivo Tests to Monitor Body Burdens of
    Toxic Metals in Man," in Chemical Toxicology and Clinical
    Chemistry of Metals, S. Brown and J. Savory, Eds. (New
    York: Academic Press, 1983), p. 113.

6. Bernard, A., and R. Lauwerys. "Effects of Cadmium in Man," in Handbook of Experimental Pharmacology: Cadmium, Vol. 80, E. C. Foulkes, Ed. (Berlin-Heidelberg-New York: Springer-Verlag, 1986), p. 135.

7. Drücke, T. "Dialysis, Osteomalacia and Aluminum Intoxication," Nephron 26:207-210 (1980).

8. Wedeen, R. P., D. K. Mallik, and V. Batuman. "Detection and Treatment of Occupational Lead Nephropathy," Arch. Int. Med. 139:53-57 (1979).

9. Lauwerys, R., H. Roels, P. Genet, G. Toussaint, A. Bouckaert, and S. De Cooman. "Fertility of Male Workers Exposed to Mercury Vapor or to Manganese Dust. A Questionnaire Study," Am. J. Ind. Med., 7:171-176 (1985).

10. Campbell, B. C., H. L. Ellitt, and P. A. Meredith. "Lead Exposure and Renal Failure: Does Renal Insufficiency Influence Lead Kinetics?" Toxicol. Letters, 9:121-124 (1981).

11. Roels, H., J. P. Gennart, R. Lauwerys, J. P. Buchet, J. Malchaire, and A. Bernard. "Surveillance of Workers Exposed to Mercury Vapor: Validation of a Previously Proposed Threshold Limit Value for Mercury Concentration in Urine," Am. J. Ind. Med. 7:47-72 (1985).

12. Vahter, M. Assessment of Human Exposure to Lead and Cadmium Through Biological Monitoring. (Stockholm: National Swedish Institute of Environmental Medicine and Karolinska Institute, 1982).

13. Lauwerys, R. H. Roels, J. P. Buchet, A. Bernard, and Ph. de Wals. "Environmental Pollution by Cadmium in Belgium and Health Damage," in Proceedings of the Third International Cadmium Conference, D. Wilson and R. Volpe, Eds. (London: Cadmium Association, 1982), p. 123.

14. Lauwerys, R., R. Hardy, M. Job, J. P. Buchet, H. Roels, P. Bruaux, and D. Rondia. "Environmental Pollution by Cadmium and Cadmium Body Burden: An Autopsy Study," Toxicol. Letters 23:287-289 (1983).

15. Wilson, H. K., S. M. Robertson, H. A. Waldron, and P. Gompertz. "Effect of Alcohol on the Kinetics of Mandelic Acid Excretion in Volunteers Exposed to Styrene Vapor," Brit. J. Ind. Med. 40:75-80 (1983).

16. Dossing, M., J. Baelum, S. M. Hansen, and G. R. Lundqvist. "Effect of Ethanol, Dimetidine, and Propranolol on Toluene Metabolism in Man," Int. Arch. Occup. Environ. Health 54:309-316 (1984).

17. Buchet, J. P., A. Geubel, S. Pauwels, P. Mahieu, and R. Lauwerys. "The Influence of Liver Diseases on the Methylation of Arsenic in Humans," Arch. Toxicol. 55:151-154 (1984).

18. Veulemans, H., and R. Masschelein. "Experimental Human Exposure to Toluene. 1. Factors Influencing the Individual Respiratory Uptake and Elimination," Int. Arch. Occup. Environ. Health 51:365-369 (1983).

19. Aitio, A., K. Pekari, and M. Järvisalo. "Skin Absorption as a Source of Error in Biological Monitoring," Scand. J. Work Environ. Health 10:317-320 (1984).

20. Lauwerys, R., H. Roels, J. P. Buchet, and A. Bernard. "Non Job Related Increased Urinary Excretion of Mercury," Int. Arch. Occup. Environ. Health 39:33-36 (1977).

21. Lauwerys, R., T. Dath, J. M. Lachapelle, J. P. Buchet, and H. Roels. "The Influence of Two Barrier Creams on the Percutaneous Absorption of m-xylene in Man," J. Occup. Med. 20:17-20 (1978).

22. Lauwerys, R., A. Kivits, M. Lhoir, P. Rigolet, D. Houbeau, J. P. Buchet, and H. Roels. "Biological Surveillance of Workers Exposed to Dimethylformamide and the Influence of Skin Protection on its Percutaneous Absorption," Int. Arch. Occup. Environ. Health 45:189-203 (1980).

CHAPTER 3

THE MONITORING OF EXPOSURE TO CARCINOGENS BY THE GC-MS
DETERMINATION OF ALKYLATED AMINO ACIDS IN HEMOGLOBIN AND OF
ALKYLATED NUCLEIC ACID BASES IN URINE

Peter B. Farmer, David E. G. Shuker, and Eric Bailey

INTRODUCTION

Exposure to alkylating carcinogens results in the covalent binding of the active genotoxic species to cellular macromolecules. Human exposure to these alkylating agents could satisfactorily be monitored by determination of the extent of this binding, ideally at the biologically significant site in deoxyribonucleic acid (DNA). However, in practice the nature of this site is not normally known with certainty, and the acquisition of sufficient carcinogen-DNA adducts for chemical determination presents considerable difficulty. For human monitoring, one is restricted to readily accessible biological media (e.g., blood) and the use of hemoglobin adducts as an indicator of the formation of carcinogen-DNA adducts has recently become established [1]. Examples will be given in this chapter of methods that we have developed, using capillary gas chromatography-mass spectrometry (GC-MS) for the determination of adducts of several simple alkylating agents (e.g., methylating, ethylating, hydroxyethylating, and hydroxypropylating) with cysteine or histidine residues in hemoglobin. Our recent work on exposure of animals to acrylamide will also be discussed. For some alkylating agents, nucleic acid adducts may be monitored by quantitation of excreted N-7-alkylated guanines. N-7-substitution of guanine (and N-3-substitution of adenine) in nucleic acids yields adducts which are unstable and which decompose to liberate the free alkylated bases. For example, the extent of the excretion of 7-alkylguanine has been shown to be directly related to exposure dose for aflatoxin B_1 [2] and

for dimethylnitrosamine [3] (liberated from the in vivo nitrosation of aminopyrine). We are currently comparing the extent
of urinary excretion of 7-alkylguanine with the amount of hemoglobin amino acid alkylation following exposure of animals to
carcinogens. Comparison of the extent of reaction of a carcinogen at different nucleophilic sites may allow predictions to
be made of its reaction at the biologically significant DNA
site, and hence of the risk associated with the exposure.

MATERIALS AND METHODS

Chemicals

    S-(2-Carboxyethyl)-L-cysteine was purchased from Fluka AG
(Fluorochem Ltd., Glossop, UK). S-(3-Amino-3-oxopropyl)-L-
cysteine was synthesized by the method of Dixit et al. [4], and
S-(3-carboxypropyl)-L-cysteine by the reaction of 4-bromo-
butyric acid with L-cystine in sodium/liquid ammonia [5]. The
chemical 7-methylguanine was purchased from Sigma Chemical Co.
(Poole, UK), and 3-methyladenine from Fluka AG.

Isolation of Alkylated Amino Acids and Alkylated Purines

    Globin was prepared from blood samples by a modification of
the method of Segerbäck et al. [6]. The protein was hydrolyzed
in 6M HCl at 110° C in vacuo, in the presence of an appropriate
amino acid internal standard. The hydrolyzate was chromato-
graphed on an ion exchange column of Dowex™ 50 H+ (AG
50W-X4) (12 x 0.8 cm), eluted with M HCl or 2M HCl, and the
fraction containing the alkylated amino acid and the internal
standard evaporated to dryness under a stream of nitrogen. The
procedure used for the isolation of urinary 7-methylguanine has
been described previously [7].

Derivatization and GC-MS

    Alkylated amino acids were esterified with 3M HCl in metha-
nol, and then acylated using heptafluorobutyric anhydride [8].
The 7-methylguanine was derivatized by heptafluorobutyroyla-
tion, followed by extractive alkylation using pentafluorobenzyl
bromide [7]. The t-butyldimethylsilyl (TBDMS) derivative of 3-
methyladenine was prepared via reaction with N-methyl-N-(tert-
butyldimethylsilyl)trifluoroacetamide in acetonitrile at 130° C

for 20 min. Derivatized samples were separated on a capillary column (25m x 0.3mm, SE52 or OV1701), housed in a Carlo Erba Mega HRGC 5160 gas chromatograph, and quantitated by multiple ion detection (MID) using a VG Analytical 70-70F double focusing mass spectrometer.

RESULTS AND DISCUSSION

The use of hemoglobin alkylation for monitoring exposure may be illustrated with the example of acrylamide. Because of its α,β-unsaturated nature, the acrylamide molecule adds readily to the SH-group of cysteine [9], yielding S-(3-amino-3-oxopropyl)-cysteine, as shown in Figure 1. Upon acidic hydrolysis this would yield S-(2-carboxyethyl)cysteine. Following intravenous administration of acrylamide to rats (50 mg/kg), we have isolated this modified amino acid from globin and have identified it mass-spectrometrically as its dimethyl ester, N-heptafluorobutyroyl derivative [EI m/z 386 (M-OCH$_3$)$^+$, 1.5%, m/z 204 (M-C$_3$F$_7$CONH$_2$)$^+$, 24.6%; CI (isobutane) m/z 418 (MH$^+$) 85.3%, m/z 386 (MH-CH$_3$OH)$^+$ 100%].

Quantitative determination of derivatized S-(2-carboxyethyl)cysteine was achieved by chemical ionization (isobutane) MID of the (M-OCH$_3$)$^+$ ion using S-(3-carboxypropyl)cysteine as internal standard. Exposure levels as low as 0.5 mg/kg can be detected. Analysis of globin from exposed animals following its enzymic hydrolysis did not show levels of carboxyethylcysteine significantly above background, supporting the belief that the adduct in the protein liberates S-carboxyethylcysteine on acidic hydrolysis, consistent with it being S-(3-amino-3-oxopropyl)cysteine. We now intend to apply

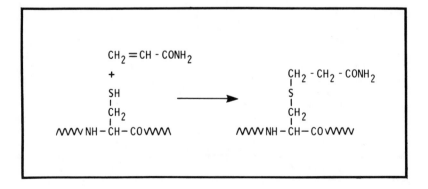

Figure 1. Reaction of acrylamide with cysteine residues in hemoglobin.

the method (using acidic protein hydrolysis) to the monitoring of human exposure to acrylamide. The background level of S-carboxyethylcysteine in human globin is not known as yet, although recent studies with rats have indicated that their globin level of S-(2-carboxyethyl)cysteine is less than 20 nmol/g protein. We are currently synthesizing acrylamide labeled with deuterium [10], with the intention of preparing from it a deuterium-labeled S-(3-amino-3-oxopropyl)hemoglobin, for application as an internal standard for acrylamide exposure monitoring. Use of such an internal standard should allow greater analytical sensitivity than the use of S-(3-carboxypropyl)cysteine as standard, as a much smaller (and hence less contaminated) amino acid fraction would need to be collected from the ion exchange separation.

The hemoglobin alkylation procedure has been used for the monitoring of human exposure, in industrial surroundings, to ethylene oxide and propylene oxide. Human propylene oxide exposure has been monitored by the quantitative determination of N^{τ}-(2-hydroxypropyl)histidine in hemoglobin using the d_5-labeled analogue of the alkylated amino acid as internal standard [8,11,12]. The homologous adduct N^{τ}-(2-hydroxyethyl)-histidine is formed following exposure to ethylene oxide [13], as shown in Figure 2, and linear dose response relationships have been observed for animals exposed to airborne concentrations of this epoxide [14]. In this case the alkylated amino acid is determined using a d_4-labeled internal standard [12] as the N,O-bis-(heptafluorobutyroyl) methyl ester derivative, as shown in Figure 3. For ethylene oxide the presence of background levels (ca. 1 nmol/g protein) of N^{τ}-(2- hydroxyethyl) histidine has limited the sensitivity for determining low exposure levels. Figure 4 shows a GC-MS calibration line obtained following the addition of varying amounts of N^{τ}-(2-hydroxyethyl)histidine, together with a fixed amount (25 ng) of the d_4-labeled internal standard, to a 10 mg sample of hydrolyzed human globin (control employee). The background level of hydroxyethylated histidine in this sample was 0.58 nmol/g globin. In a major study by Van Sittert et al. [15], no difference was observed in the histidine hydroxyethylation levels between a control population and a population occupationally exposed to low levels of ethylene oxide. However, in a recent limited study of ours [5], we have seen evidence for a dose-related increase in N^{τ}-(2-hydroxyethyl)histidine, which was confirmed by independent determination of hydroxyethylation by the measurement of the N-terminal N-(2-hydroxyethyl)valine levels.

The lifetime of hemoglobin alkylation adducts may in some cases approach the lifetime of the protein, and thus their determination represents an integral of carcinogen-dose received over this period. In contrast, the determination of urinary purine alkylation adducts is more suited for the monitoring of acute exposure, as the excretion is complete within around 5 days of the exposure [3]. Again, the presence of background levels of alkylated purines may limit the sensitivity of the

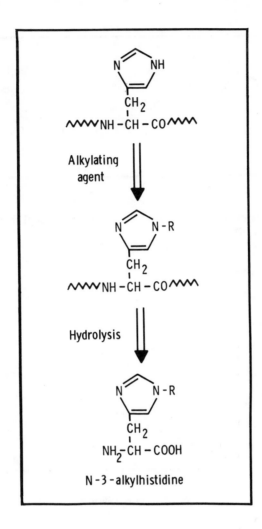

Figure 2. Reaction of alkylating agents with histidine in hemoglobin. For ethylene oxide the alkyl group R is
$CH_2 CH_2OH$.

assay (e.g., for 7-methylguanine). For this reason, our studies of methylating carcinogens have used stable isotope-labeled
analogues of the carcinogens.

In this way we have found that the ratio of N-methylation
of guanine to S-methylation of hemoglobin cysteine varies according to the methylating agent used, i.e., an S_N1 agent
dimethylnitrosamine yields relatively more 7-methylguanine than

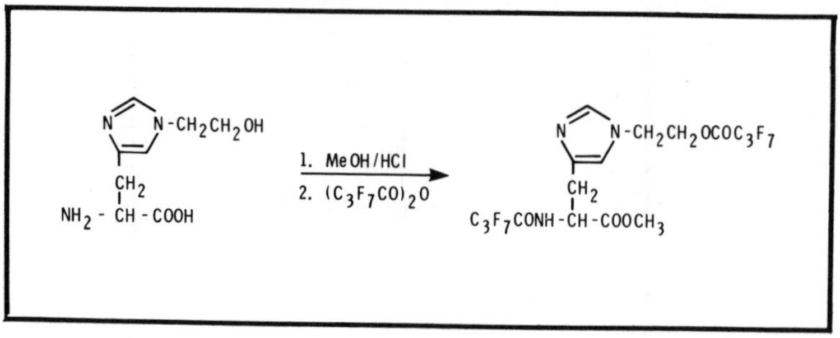

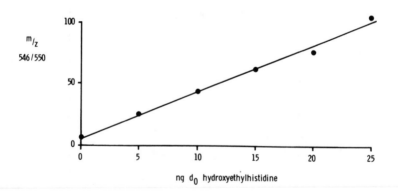

Figure 3. Derivatization of N^{τ}-(2-hydroxyethyl)histidine for GC-MS.

Figure 4. GC-MS calibration line for N^{τ}-(2-hydroxyethyl) histidine in human globin. Samples (10 mg) of hydrolyzed protein were spiked with d_0-N^{τ} (2-hydroxyethyl)histidine (0-25 ng) and d_4-N-$^{\tau}$ (2-hydroxyethyl)histidine (25 ng). After ion exchange purification and derivatization of the samples, the ions m/z 546 (d_0) (M-COOCH$_3$)$^+$ and m/z 550 (d_4) (M-COOCH$_3$)$^+$ were monitored.

an S_N2 agent methyl methanesulfonate [16]. (These experiments were carried out using d_3-methyl methanesulfonate and d_6-dimethylnitrosamine, liberated from in vivo nitrosation of d_6-aminopyrine [7]. $S-CD_3$-cysteine and $N-7-CD_3$-guanine were determined by GC-mass spectral MID.) Similarly, d_3-N-methyl-N-nitrosourea, another S_N1 agent, yielded a high $7-CD_3$-guanine to $S-CD_3$-cysteine ratio. Comparison of results from alkylated hemoglobin and urinary alkylated purines may thus give information regarding the mechanism of the alkylation reaction, and hence of the risk associated with the exposure.

Analytical methods for N-3-methyladenine are currently being developed, as there is no published evidence for background levels of this purine in urine. If no background levels of N-3-methyladenine exist, this is a much more sensitive approach to the monitoring of exposure to methylating carcinogens. The 3-methyladenine may be separated from urine by XAD-2 column chromatography, followed by reverse phase HPLC, and then determined by GC-MS MID. The TBDMS derivative of 3-methyladenine appears to have satisfactory GC and MS properties for quantitative determinations [EI m/z 263 (M+), 1.8%, m/z 206 $(M-C_4H_9)^+$, 100%; CI (isobutane) m/z 264 (MH+), 100%, m/z 206 $(M-C_4H_9)^+$, 73.9%]. Similarly, isolation procedures for urinary N-7-(2-hydroxyethyl)guanine are being developed. This alkylated base, which may be of value for monitoring acute ethylene or ethylene oxide exposures, may be isolated from urine by Dowex 50H+ ion exchange chromatography and/or HPLC (C-18 column, eluting solvent 0.1% heptafluorobutyric acid/methanol).

At present all alkylated bases that we have studied are analyzed as derivatives by capillary GC-MS, following their partial purification from the urine. An alternative approach that we are now considering is the use of a triple sector mass spectrometer with analysis of collision spectra ions. As an example of this technique, Figure 5(b) shows a daughter ion spectrum (collision gas air, collision energy 13eV), (obtained with kind permission of VG Analytical Ltd. on a 7070 EQ mass spectrometer), of 7-methylguanine. Ions of mass 165 (M+ for 7-methylguanine) were collected by the electric and magnetic sectors, fragmented in a collision cell, and the resulting ions focused in the final quadrupole sector. Figure 5(a) illustrates the spectrum obtained in a similar fashion when rat urine (3 1) was placed on the MS probe, and clearly shows the potential of the method for detecting alkylated products in urine. The background level of 7-methylguanine that is being seen in Figure 5(a) is ca. 10 µg/ml.

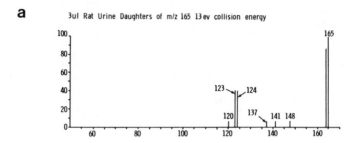

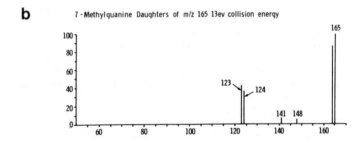

Figure 5. (a)   Rat urine daughter ion scan. (Collision gas
air, collision energy 13 eV, EI, m/$_z$ 165, VG 7070-
EQ instrument.)

(b) 7-Methylguanine daughter ion scan. Conditions
as in (a). (Spectra obtained by VG Analytical Ltd.,
Wythenshawe, Manchester, England).

CONCLUSION

As illustrated above, exposure of animals or man to alkylating carcinogens may be monitored by determinations of alkylated proteins or of alkylated nucleic acid bases in urine. For
the chemically simple alkylating agents that we have studied
(containing up to 3 carbon atoms) the presence of background
levels of alkylated products may limit the sensitivity of this
approach for exposure monitoring. The source of these backgrounds is of particular interest to investigate, in case they
are caused by unsuspected carcinogenic hazards.

We would predict that as the alkyl group becomes more complex in the carcinogenic species, the abundance of "background
alkylations" would decrease. For the sensitive and specific
analysis of the adducts resuting from these more complex carcinogenic molecules, adaptation of the analytical methods will

be required. In particular, the work-up procedures should be modified in order to separate the alkylated adduct (or the alkyl function itself) more effectively from the normal protein or nucleic acid constituents. Two recently published techniques which may be of particular importance are the analysis of N-terminal valine adducts in hemoglobin by a modified Edman procedure [17], and the analysis of exposure to aromatic amines following their hydrolytic release from their adducts with hemoglobin cysteine [18]. Analytical developments, such as MS-MS and HPLC-MS, may also increase the range and specificity of the exposure-monitoring procedures.

DISCLAIMER

The work described in this chapter was not funded by EPA and no official endorsement should be inferred.

REFERENCES

1. Ehrenberg, L. and S. Osterman-Golkar. "Alkylation of Macromolecules for Detecting Mutagenic Agents," Teratog. Carcinog. Mutagen. 1:105-127 (1980).

2. Bennett, R. A., J. M. Essigmann, and G. N. Wogan. "Excretion of an Aflatoxin-Guanine Adduct in the Urine of Aflatoxin B_1-treated Rats," Cancer Res. 41:650-654 (1981).

3. Shuker, D. E. G., E. Bailey, and P. B. Farmer. "Methylation of Proteins and Nucleic Acids in Vivo. Use of Trideuteromethylating Agents or Precursors," in Proceedings of the Eighth International Meeting on N-Nitroso Compounds, IARC Scientific Publication No. 57 (1984), pp. 589-594.

4. Dixit, D., P. K. Seth, and H. Mukhtar. "Metabolism of Acrylamide into Urinary Mercapturic Acid and Cysteine Conjugates in Rats," Drug. Metab. Dispos. 10:196-197 (1982).

5. Farmer, P. B. Unpublished results (1985).

6. Segerbäck, D., C. J. Calleman, L. Ehrenberg, G. Lofroth, and S. Osterman-Golkar. "Evaluation of Genetic Risks of Alkylating Agents. IV. Quantitative Determination of Alkylated Amino Acids in Haemoglobin as a Measure of the Dose after Treatment of Mice with Methyl Methanesulfonate," Mutation Res. 49:71-82 (1978).

7. Shuker, D. E. G., E. Bailey, S. M. Gorf, J. Lamb, and P. B. Farmer. "Determination of N-7-[^{2}H$_3$]Methylguanine in Rat Urine by Gas Chromatography Mass Spectrometry Following Administration of Trideuteromethylating Agents or Precursors," Anal. Biochem. 140:270-275 (1984).

8. Farmer, P. B., S. M. Gorf, and E. Bailey. "Determination of Hydroxypropylhistine in Haemoglobin as a Measure of Exposure to Propylene Oxide using High Resolution Gas Chromatography Mass Spectrometry," Biomed. Mass Spectrom. 9:69-71 (1982).

9. Hashimoto, K., and W. N. Aldridge. "Biochemical Studies on Acrylamide, a Neurotoxic Agent," Biochem. Pharmacol. 19:2591-2604 (1970).

10. Farmer, P. B., I. Bird, E. Bailey, and D. E. G. Shuker. "The Use of Deuterium Labelling in Studies of Protein and DNA Alkylation," in Proceedings of the Second International Symposium on the Synthesis and Applications of Isotopically Labelled Compounds, in press (1985).

11. Osterman-Golkar, S., E. Bailey, P. B. Farmer, S. M. Gorf, and J. H. Lamb. "Monitoring Exposure to Propylene Oxide Through the Determination of Haemoglobin Alkylation," Scand. J. Work Environ. Health 10:99-102 (1984).

12. Campbell, J. B. "The Synthesis of N(τ)-(2-Hydroxypropyl) Histidine, N(τ)-(2-Hydroxyethyl) Histidine and their Deuterated Analogues," J. Chem. Soc. Perkin Trans. 1:1213-1217 (1983).

13. Osterman-Golkar, S., L. Ehrenberg, D. Segerbäck, and I. Hallstrom. "Evaluation of Genetic Risks of Alkylating Agents. II. Haemoglobin as a Dose Monitor," Mutation Res. 34:1-16 (1976).

14. Osterman-Golkar, S., P. B. Farmer, D. Segerbäck, E. Bailey, C. J. Calleman, K. Svensson, and L. Ehrenberg. "Dosimetry of Ethylene Oxide in the Rat by Quantitation of Alkylated Histidine in Hemoglobin," Teratog. Carcinog. and Mutagen. 3:395-405.

15. Van Sittert, N. J., G. DeJong, M. G. Clare, R. Davies, B. J. Dean, L. J. Wren, and A. S. Wright. "Cytogenetic, Immunological and Haematological Effects in Workers in an Ethylene Oxide Manufacturing Plant," Brit. J. Indust. Med. 42:19-26 (1985).

16. Farmer, P. B., D. E. G. Shuker, and I. Bird. "DNA and Protein Adducts as Indicators of In Vivo Methylation by Nitrosatable Drugs," Carcinogenesis 7:49-52 (1986).

17. Törnqvist, M., J. Mowrer, S. Jensen, and L. Ehrenberg. "Monitoring of Environmental Cancer Initiators through Hemoglobin Adducts by a Modified Edman Degradation Method," Anal. Biochem., in press (1986).

18. Green, L. C., P. L. Skipper, R. J. Turesky, M. S. Bryant, and S. R. Tannenbaum. "In Vivo Dosimetry of 4-Aminobiphenyl in Rats via a Cysteine Adduct in Hemoglobin," Cancer Res. 44:4254-4259 (1984).

CHAPTER 4

DETERMINING DNA ADDUCTS BY ELECTROPHORE LABELING-GC

Roger W. Giese

INTRODUCTION

We have begun to work on the determination of deoxyribonucleic acid (DNA) adducts in biological samples by electrophore labeling gas chromatography (GC). Our overall analytical
strategy consists basically of the following steps: (1) iso-
late the DNA from the biological sample by conventional techniques; (2) hydrolyze the DNA to bases or nucleosides; (3) iso-
late the DNA adducts from the bulk of normal DNA components;
(4) label the adducts with an electrophore; and (5) use GC with
electron capture detection (ECD) or detection by negative ion
chemical ionization mass spectrometry (NCI-MS) to quantify the
adducts with high sensitivity.

In this chapter, we will provide an overview of our work to
date on the use of electrophore labeling-GC for measuring DNA
adducts. First, we will discuss the nature and role of the
electrophore labels that we are using to make the adducts highly sensitive. Next, our plans for sample cleanup will be presented with emphasis on the above step (3). This step is likely to be more challenging than the earlier steps (1) and (2) of
sample cleanup because less work has been done previously in
step (3). In this same part of our discussion we will point
out the key role that is anticipated for HPLC and immunoaffinity chromatography.

Most of our actual work to date has been concerned with
step (4), in which the DNA adduct is electrophore-labeled.
Little work previously has been done on the attachment of electrophores to nucleobases and nucleosides. One of the key questions at the outset of our work two years ago was whether suitable electrophoric derivatives of DNA adducts could be prepared

for ultratrace GC analysis. This includes concerns for the hydrolytic and thermal stability of these derivatives, their yields, and their GC-ECD/NCI-MS characteristics. We are pleased to report that such derivatives indeed can be prepared, at least for the adducts and model adducts that we have investigated to date involving pyrimidines. This is all presented here in three sections discussing criteria for electrophoric derivatives, electrophore labeling of pyrimidine bases, and our work on O^4-ethylthymidine.

We will then discuss our strategy for the use of an internal standard in this project. Although an istopically labeled form of the analyte is the most reliable internal standard, it may sometimes be more convenient and acceptable to use an internal standard in which the structural variation is incorporated into a derivatizing group.

Next, we will explain the potential use of an indirect class of electrophores called "release tags" for determining more complex DNA adducts. Release tags potentially allow some of the advantages of electrophores to be applied to DNA adducts that are labile or too large for direct analysis by GC techniques. Finally, we briefly define the general advantages and disadvantages of electrophore labeling-GC for measuring DNA adducts.

ELECTROPHORES

An electrophore is a molecule that captures a low-energy electron in the gas phase. The immediate consequence of this capture is the formation of an anion radical. This event takes place and can be detected in both an electron capture detector (ECD) and in a negative chemical ionization source of a mass spectrometer (MS). In the ECD the loss of the thermal electron is detected, while the MS detects the anion radical or a subsequent anionic fragment derived from this radical.

While electrophores show a propensity relative to "ordinary molecules" to capture a thermal electron, electrophores range in this property from weak to strong. Much work has been done on the relative strengths and therefore ease of detection of electrophores by ECD or NCI-MS [e.g., 1], but only guidelines rather than exact rules exist for predicting electrophic properties of novel structures. It is important from a practical standpoint to identify molecular structural features yielding strong electrophores because such compounds, by definition, are detected with highest sensitivity by ECD and NIC-MS. They therefore will tend to give the most sensitive derivatives when attached to DNA adducts.

Polyhalogenated organic compounds like lindane and carbon tetrachloride are common examples of strong electrophores. For derivatization purposes, however, a functional group must be available. Thus electrophoric derivatizing reagents such as pentafluorobenzoyl chloride, pentafluorobenzyl bromide, and heptafluorobutyric anhydride are commonly used to form electrophoric derivatives of analytes.

SAMPLE CLEANUP

Fortunately, DNA is a relatively unique macromolecule in biological samples, allowing its convenient purification from other components in these samples. Solvent extraction, precipitation, and ion exchange chromatographic steps are frequently used. Thus, once electrophore-labeling methodology becomes successful for standards of DNA spiked with ultratrace amounts of authentic adducts, it should generally be a straightforward process to extend the methodology to biological samples from exposed individuals.

The isolated DNA can then be acid- or nuclease-hydrolyzed, yielding the adducts as base or nucleoside products in most cases. Probably high-performance liquid chromatography (HPLC) or an immunoaffinity column will then generally be employed to fish out the adducts from the large background of normal bases and nucleosides prior to electrophore labeling of these adducts for subsequent determination by GC-ECD/NCI-MS. HPLC is well-established as a high resolution technique for resolving similar bases or nucleosides, as has been reviewed [2]. Sample cleanup by immunoaffinity chromatography prior to DNA adduct detection by radioimmunoassay (RIA) has been demonstrated by Groopman et al. [3]. Immunoaffinity chromatography also has been used for other classes of trace analytes, e.g., in the determination of angiotensin II in serum by radioactive labeling [4].

Base adducts will generally be preferred, due to their simpler structures and higher volatility characteristics for direct electrophore labeling. More volatile products can generally be determined with higher sensitivity both by GC-ECD and GC-NCI-MS. However, some adducts will not survive the hydrolysis conditions required to yield the bases. Nucleoside adducts are attractive because they can be obtained by enzymatic hydrolysis, and in each case a common functional group, the sugar, is available for labeling. Nevertheless, nucleosides possess a greater variety of functional groups and the glycosidic bond becomes labile for certain modifications of the base [5].

CRITERIA FOR ELECTROPHORIC DERIVATIVES

Ideally, the DNA adduct is reacted with a strong electrophore to form a single product that is hydrolytically stable

and has favorable detection properties by GC-ECD/NIC-MS. However, there are many pitfalls for this important stage of the
analysis. The complex chemical nature of DNA adducts makes it
challenging to avoid side products. More than one derivatization reaction may be necessary to fully remove active hydrogens. Reactions that give a reasonable yield of a desired product at a conventional level of adduct (e.g., mg amount) may
experience difficulties when applied to a sub-ng amount. A
product that is successful for determination by GC-ECD may not
give an analyte-specific ion by NCI-MS. Instead, an anionic
fragment corresponding to only the strong electrophore may be
seen by the latter technique. Thus, there are many challenges
for the electrophore-labeling stage of determining DNA adducts
by GC-ECD/NCI-MS. In fact, clearly this is the stage which
needs to be addressed first in developing this methodology, a
task that we have undertaken.

ELECTROPHORE LABELING OF PYRIMIDINE BASES

    At the outset of our work two years ago, little had been
done on the reaction of electrophores with nucleobases and nucleosides. This is in spite of an extensive amount of work on
the analysis of such substances by GC, including GC-MS, as has
been reviewed [6]. Strong electrophores had not been attached
to nucleobases or nucleosides at all. The entire electrophore
literature for nucleic-acid products consisted of only three
articles. Geligkens et al. [7], in an important paper, reported the attachment of a trifluoroacetyl electrophore to two of
the DNA bases, cytosine and guanine, followed by peralkyla-
tion. They obtained good products for GC analysis, and their
overall methodology was applied to both standards and DNA samples. Although detection was done by electron impact MS, the
potential for using ECD and NCI-MS to optimize the sensitivity
was pointed out.
    Two reports have appeared in which GC-ECD has been applied
to the analysis of DNA bases or analogous substances. In the
first, thymine was quantified after derivatization with 1,3-
bis(chloromethyl)tetramethyldisilazane for the microdetermination of DNA in biological samples [8]. Trifluoroacetylated derivatives of cytokinins were analyzed in the second case [9].
Detection limits reached the low picogram level in both of
these studies.
    We began investigating the usefulness of GC techniques for
the ultratrace determination of DNA adducts by reacting the two
strong electrophores, pentafluorophenylsulfonyl chloride (PPSC)
and pentafluorobenzoyl chloride (PFBC), with pyrimidine bases.
PPSC had been recently introduced as a reagent for forming
stable derivatives of tyrosyl peptides for determination with
high sensitivity by GC-ECD [10]. PFBC was a logical choice for
an acylating reagent, because it tends to form more strongly

electrophoric derivatives with amines than does trifluoroacetic
anhydride [11]. Also, PFB-amines are more hydrolytically
stable than their trifluoroacetyl counterparts [12].

We consider that GC-ECD/NCI-MS techniques will be fully
successful for the ultratrace determination of DNA adducts only
if electrophoric derivatives can be obtained that are hydrolytically stable. This is because the demands of such analysis
will certainly require some sample cleanup after derivatization, exposing these derivatives to traces or even bulk amounts
of water. This includes the likely need for more than one derivatization reaction to deal with the structural complexity of
nucleobases and nucleosides, with accompanying intermediate
extraction and evaporation steps.

Thus, the first test that we applied to our electrophoric
derivatives of the bases cytosine, thymine, and uracil, shown
in Figure 1, was their aqueous stability. We were pleased to
see, as shown in Table 1, that the aqueous stabilities of these
derivatives ranged from good to excellent, especially considering the significant hydrolytic stress that we applied. We
kept them in water for 7 hours under both acidic (acetate
buffer, pH 4) and nucleophilic basic (Tris, pH 8) conditions.

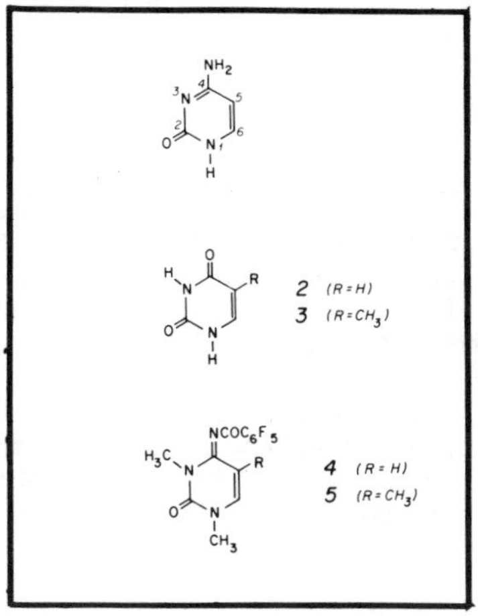

Figure 1. Structures of cytosine (1); uracil (2); thymine
(3); N4-PFB-1,3-dimethylcytosine (4); N4-
PFB-1,3-dimethyl-5-methylcytosine (5). Reprinted
with permission from A. Nazareth, M. Joppich, S.
Abdel Baky, K. O'Connell, A. Sentissi, and R. W.
Giese. "Electrophore-Labeling and Alkylation of
Standards of Nucleic Acid Pyrimidine Bases for
Analysis by Gas Chromatography with Electron
Capture Detection," J. Chromatogr. 314:201-210
(1984). Copyright 1984 Elsevier Science Publishers
B.V. Amsterdam.

Table 1. Stability of Electrophore-Labeled Nucleic Acid
          Pyrimidine Bases.[a]
The buffers were: acetate (0.1 M sodium acetate, pH 4);
ACN-Phos. [acetonitrile-0.001 M sodium phosphate, pH 5 (55:45,
v/v)] and Tris [0.2 M tris(hydroxymethyl)aminomethane].

| | Recovery After 7 h (%) | | |
Compound	Acetate, pH 4	ACN-Phos., pH 5 (55:45)	Tris, pH 8
Cytosine			
N^4-HFB-1,3-dimethyl-	86	100	0
N^4-PFB-1,3-dimethyl-	100	100	100
N-PPS-*	61	84	59
Thymine			
PFB-*	100	100	7
PPS-methyl	100	100	64
Uracil			
PFB-*	75	94	19
PPS-methyl-	100	100	36

[a]Reprinted with permission from A. Nazareth, M. Joppich, S.
Abdel-Baky, K. O'Connell, A. Sentissi, and R. W. Giese.
"Electrophore-Labeling and Alkylation of Standards of Nucleic
Acid Pyrimidine Bases for Analysis by Gas Chromatography with
Electron Capture Detection," J. Chromatogr. 314:201-210 (1984).
Copyright 1984 Elsevier Science Publishers B.V.
*Methylated derivatives of PPS-cytosine, PFB-thymine, and
PFB-uracil are not reported due to the instability of these
starting materials to our alkylation conditions.

Also, we dissolved them in a typical HPLC mobile phase (55%
acetonitrile:45% pH 5 phosphate). The stability of a hepta-
fluorobutyryl (HFB) derivative of cytosine was also determined
for comparison purposes, as shown in Table 1. While this
latter derivative can tolerate mildly acidic aqueous conditions
fairly well, it is fully hydrolyzed in the Tris buffer.
    The responses by GC-ECD for these derivatives were in the
vicinity of that of lindane, a strong electrophore. For the
N^4-1,3-dimethyl derivative of cytosine, a detection limit at
the low fg level was seen both by GC-ECD [13] and GC-NCI-MS
[14]. For the latter determination, the base peak was the molecular ion. The detection of 1 fg (3 x 10^{-18} mole) of compounds 4 and 5 by GC-NCI-MS is shown in Figure 2.
    Encouraged by these results, we are continuing to pursue
the determination of 5-methylcytosine as a model DNA adduct by

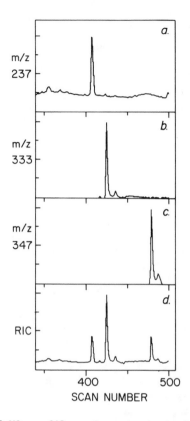

Figure 2. GC-NCI-MS profiles of a standard mixture of 1 fg of
derivatives 4 and 5 from Figure 1 with hepta-
chlor as the internal standard. Bottom trace (d)
represents the reconstructed total ion current chromatogram. Single ion profiles of the internal
standard, and compounds 4 and 5 are traces a, b,
and c, respectively. Reprinted with permission from
G. B. Mohamed, A. Nazareth, M. J. Hayes, R. W.
Giese, and P. Vouros. "GC-MS Characteristics of
Methylated Perfluoroacyl Derivatives of Cytosine and
5-Methyl Cytosine," J. Chromatogr. 314:211-217
(1984). Copyright 1984 Elsevier Science Publishers
B. V. Amsterdam.

electrophore labeling-GC. The next step is to extend the derivatization reaction to a small amount of this analyte. This
work is best studied first by HPLC. The cytosine derivative
can be detected down to the low ng level by this technique.
The advantage of using HPLC is that the reaction steps can be
monitored at an intermediate level of derivatization where active hydrogens are still present so that GC cannot be used.

Also, unreacted starting material can be determined along with any side products that fail to elute by GC. Our initial results extending our electrophore derivatization reaction of cytosine to lower levels via HPLC monitoring have been reported [15]. Starting with 50 nmol of cytosine, the overall yield of product is $59 \pm 4.6\%$. This is not an unreasonable result, but more recently we are pursuing a pivalyl, pentafluorobenzyl derivative of this base that is giving a higher yield even when applied to 150 pg of 5-methylcytosine derived from a hydrolyzate of calf-thymus DNA [16].

O^4-ETHYLTHYMIDINE

We have also begun to explore electrophore labeling and GC of nucleosides. A key advantage of the latter as a form for DNA adducts is the inability of some adducts to survive the stronger hydrolytic conditions necessary to degrade DNA down to bases.

We have found that O^4-ethylthymidine and some related nucleosides, along with the base thymine, can be derivatized with pentafluorobenzyl bromide using phase transfer alkylation conditions [17]. All active hydrogens, both on the base and sugar, are alkylated in this compound. The structure of the derivative for O^4-ethylthymidine is shown in Figure 3. Table 2 shows the compounds that we derivatized, along with their molecular weights and relative molar responses. As with the

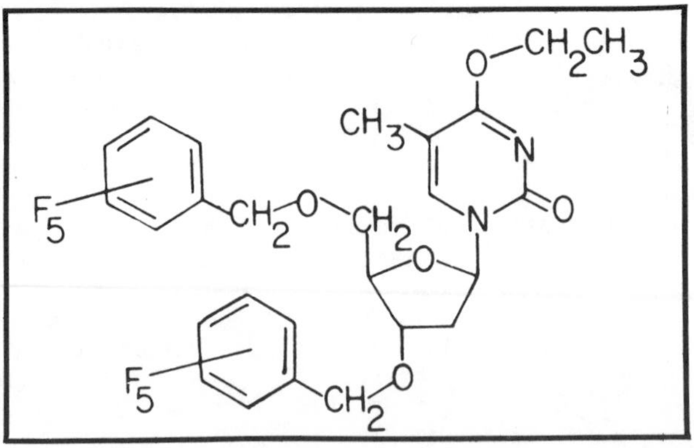

Figure 3. Structure of $3',5'$-bis-(O-pentafluorobenzyl)-O^4-ethylthymidine (compound 4 in Table 2 and Figure 4). Reprinted with permission from J. Adams, M. David, and R. W. Giese. "Pentafluorobenzylation of O^4-Ethylthymidine and Analogs by Phase-Transfer Catalysis for Determination by Gas Chromatography with Electron Capture Detection," Anal. Chem., 58: 345-348 (1986). Copyright 1986 American Chemical Society.

Table 2. GC-ECD Characteristics of the Pentafluorobenzyl
(PFBz) Derivatives.[a]

Compound	No.[b]	Mol Wt.	Rel Molar Response[c]
1,3-bis(PFBz)thymine	2	486	1.6 $\pm$ 0.11
3',5'-bis-($\underline{O}$-PFBz)-			
3-methylthymidine	3	616	1.1 $\pm$ 0.062
O^4-ethylthymidine	4	630	0.60 $\pm$ 0.068
3-(PFBz)thymidine	5	782	1.5 $\pm$ 0.092

[a]Reprinted with permission from J. Adams, M. David, and R.
W. Giese. "Pentafluorobenzylation of O^4-Ethylthymidine and
Analogs by Phase-Transfer Catalysis for Determination by Gas
Chromatography with Electron Capture Detection," Anal. Chem.,
58:345-348 (1986). Copyright 1986 American Chemical Society.
[b]Refers to peak number in Figure 4.
[c]Area units/mol relative to lindane; represents mean +
standard deviation from 1 to 2 injections of duplicate sets
of dilutions at each concentration level containing all four
compounds and lindane covering the linear range: for compound
2, $\underline{n}$ = 18; for 3, $\underline{n}$ = 26; for 4, $\underline{n}$ = 42, for 5, $\underline{n}$ = 26. ($\underline{n}$ =
total number of data points throughout the linear range.)

pyrimidine bases discussed above, these nucleoside derivatives
are seen to have responses near that of lindane, a strong electrophore.

The good performance of these compounds when determined by
GC-ECD is shown in Figure 4. Minimal tailing is seen for the
peaks that are also well-resolved. For derivatized O^4-
ethylthymidine, the detection limit, shown in Figure 4C, is 27
fg (4.5 x 10^{-17} mol). This extends the detection limit for
nucleoside GC by 10^3.

Currently, we are extending this derivatization procedure
for O^4-ethylthymidine to smaller amounts of this analyte,
with monitoring of the reaction by HPLC according to the guidelines presented above for the derivatization of 5-methylcyto-
sine. Phase transfer alkylation with pentafluorobenzyl bromide
is an attractive derivatization reaction because it involves
relatively mild conditions and removes all active hydrogens in
a single step for the compounds investigated here. We expect
that many other DNA adducts will be converted to appropriate
derivatives for GC-ECD/NCI-MS by this technique. For example,
we have found that the adduct 5-hydroxymethyluracil can be successfully derivatized by phase transfer alkylation with pentafluorobenzyl bromide [18].

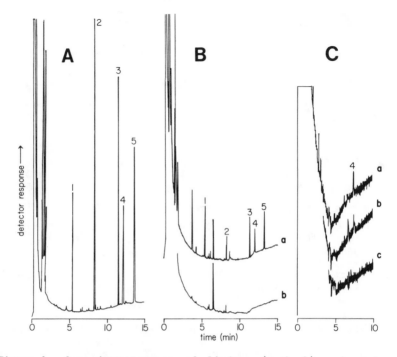

Figure 4.   Gas chromatograms of lindane (peak 1) and pentafluorobenzyl derivatives. Peaks 2-5 refer to derivatives listed in Table 2.   One µl of analytical
standards containing a mixture of all five compounds
in toluene was injected.   Chromatogram A: 1=1.1,
2=4.0, 3=5.0, 4=5.0, and 5=6.0 pg; attenuation = 64
x 1.   Chromatogram B: (a) 1=0.11 pg, 2=0.53, 3=0.15,
4=0.28, and 5=0.20 pg; (b) = blank; attenuation = 16
x 1.   Chromatogram C: (a) 57 fg and (b) 27 fg
(0.045 fmol) of compound 4; (c) = blank; attenuation = 4 x 1.   Reprinted with permission from J.
Adams, M. David, and R. W. Giese.   "Pentafluorobenzylation of O^4-Ethylthymidine and Analogs by
Phase-Transfer Catalysis for Determination by Gas
Chromatography with Electron Capture Detection,"
Anal. Chem., 58: 345-348 (1986).   Copyright 1986
American Chemical Society.

INTERNAL STANDARD

An important advantage of GC techniques for determining DNA
adducts relative to other approaches, such as ^{32}P-post labeling TLC analysis [19] and immunoassay [20], is the ease with
which GC methodology can incorporate an internal standard to

enhance both the accuracy and precision. However, there is no best internal standard when both practical and theoretical considerations are taken into account. For example, although an appropriate stable isotope derivative of the analyte for determination by GC-MS is the most powerful approach, obtaining this type of internal standard for an analyte can be an expensive undertaking when many analytes are to be determined. We will apply it here as necessary, but are also interested in testing another approach in which the internal standard is a close analog of the derivatized analyte. This internal standard will be prepared by derivatizing an authentic sample of the analyte with a slightly different chemical group than is used in the analytical procedure. An example is our use (work in progress) of a tetrafluorobenzyl derivative of 5-methylcytosine as an internal standard in the procedure to determine 5-methylcytosine by labeling with a pentafluorobenzyl group. This strategy makes internal standards readily available, and will monitor the analytical sequence after the derivatization step. The price for this, unfortunately, is that the pre-derivatization and derivatization steps are not intrinsically monitored. We believe, however, that satisfactory monitoring of these steps can still be done. First of all, external standardization can be used. Second, an analog of the analyte present at the ng level can be added to all of the samples to "lock in" the performance of the derivatization reaction. The derivatization reaction can thus be monitored by subjecting control reaction samples containing a sufficient amount of this analog for HPLC analysis. In this manner, the yield of the derivatization reaction can be defined for each batch of samples. Finally, the method of standard additions is available.

RELEASE-TAGS GC

It would seem that a major disadvantage of GC methodology is its limitation to the analysis of volatile, thermally stable compounds or derivatives. While this is true for direct analysis by GC, in which the observed peak has a retention time characteristic of the analyte, it is not true for analysis by "release-tag GC" [21]. In GC with a release-tag electrophore, the electrophore is first attached to the analyte through a cleavable linkage such as a methionylamide [22], glycol [23], or olefin [23] group. The analyte, after undergoing thorough purification, is then determined indirectly by chemical release to liberate the electrophore for quantitation by GC. Thus, this approach utilizes the principle of quantitative analysis by isotope derivatization [24]. An example is the determination of thyroxine in serum by release-tag GC analysis [22].

The qualitative power of release tag electrophore labeling GC is weaker than analysis by labeling with a direct electrophore, but should be comparable to that of ^{32}P-post labeling TLC technology. Release-tag GC avoids the handling problems of

^{32}P, and the limited ability of the ^{32}P technique to incorporate a good internal standard. Although ^{32}P-post labeling has an advantage of utilizing the specificity of an enzyme in the labeling step, different adducts may vary in the ease of their labeling by an enzyme, the product is a nucleotide that can be difficult to purify, and one is limited to a single labeling reaction. The high cost of ^{32}P also prevents it from being used other than in a radioenzymatic procedure. In contrast, many reaction techniques can be employed with release tags, and the products, not being radioactive or ionic, can be subjected to physical (e.g., HPLC) and other purification characterization steps, particularly when monitored by an internal standard that is a close structural analog of the analyte.

Thus, release-tag electrophores can be investigated for more complex DNA adducts that fail to yield volatile, thermally stable derivatives for direct analysis by GC. For example, there is some interest in putative DNA-protein adducts arising from exposure to formaldehyde [25]. Even after potential enzymatic digestion of such adducts down to a nucleoside-amino acid product, it is likely that such adducts would still be difficult to directly quantify at an ultratrace level, including derivatization, by GC. Release-tag GC then provides an alternative approach.

ADVANTAGES

Electrophore labeling-GC-MS methodology can greatly advance the determination of DNA adducts. It is ultrasensitive, definitive, avoids handling of radioisotopes, can potentially determine several adducts simultaneously, utilizes internal standards, and can discover and elucidate the structures of unknown adducts. Alternate immunoassay and ^{32}P-post labeling methodologies, although successful for determining DNA adducts, do not provide some of these advantages. For example, there is little structural information in the autoradiographic spots that are the final outcome of determining DNA adducts by ^{32}P post labeling. Immunoassays tend to require a separate, high-affinity and high-specificity antibody for each adduct. Such an antibody may not always be available, or may require considerable time to develop, whereas electrophore-GC-MS methodology potentially can move quickly onto new adducts once the general methodology has been developed.

DISADVANTAGES

Electrophore labeling-GC-MS methodology lacks the convenience of established immunoassays (although immunoassays for ultratrace analytes are less convenient than those for routine analytes), is less proved at the present time than successful ^{32}P-post labeling methodology (started several years ahead of electrophore-labeling methodology for this application), utilizes expensive instrumentation, and involves the difficulties of chemical derivatization of ultratrace analytes.

CONCLUSION

Much work remains to be done in bringing electrophore labeling-GC methodology to the determination of DNA adducts in biological samples. If the goal were merely to quantify μg or ng amounts of these adducts, then the methodology currently available would be fairly satisfactory. However, the goal is to quantify such adducts in amounts significantly below the ng level. This greatly increases the difficulty for most of the steps involved. Nevertheless, given the good performance of the electrophoric derivatives obtained to date for some initial DNA adducts and model adducts, the current availability of powerful sample cleanup steps such as HPLC and immunoaffinity chromatography, and the high resolution available from GC-ECD and especially GC-NCI/MS, it is clear that electrophore labeling-GC will play a major role in the determination of DNA adducts.

ACKNOWLEDGMENTS

The work described in this chapter has been funded in part by grant CR812740 from the Reproductive Effects Assessment Group of the United States Environmental Protection Agency (EPA), and National Cancer Institute (NCI) grant CA35843 to Northeastern University. This is contribution no. 286 from the Barnett Institute of Chemical Analysis.

REFERENCES

1. Corkill, J. A., M. Joppich, S. H. Kuttab, and R. W. Giese. "Attogram Level Detection and Relative Sensitivity of Strong Electrophores by Gas Chromatography with Electron Capture Detection," Anal. Biochem., 54:481-485 (1982).

2. Brown, P. R., A. M. Krstulovic, and R. A. Hartwick. "Current State of the Art in HPLC Analyses of Free Nucleotides, Nucleosides and Bases in Biological Fluids," in Advances in Chromatography, Vol. 18, J. C. Giddings, E.

Grushka, J. Cazes, and P. R. Brown, Eds. (New York: Marcel Dekker, Inc., 1980), Chapter 3.

3. Groopman, J. D., P. R. Donahue, J. Zhu, J. Chein, and G. N. Wogan. "Aflatoxin Metabolism in Humans: Detection of Metabolites and Nucleic Acid Adducts in Urine by Affinity Chromatography," Proc. Natl. Acad. Sci. USA 82:6492-6496 (1985).

4. Emanuel, R. L., R. Joppich-Kuhn, G. H. Williams, and R. W. Giese. "Studies Directed Toward Labeling Analysis of Angiotensin II in Plasma," Clin. Chem. 31:1723-1728 (1985).

5. Singer, B., and D. Grunberger. Molecular Biology of Mutagens and Carcinogens. (New York: Plenum Press, 1983).

6. Burlingame, A. L., K. Straub, and T. A. Baillie. "Mass Spectrometric Studies on the Molecular Basis of Xenobiotic-induced Toxicities," Mass Spectrom. Revs. 2:331-387 (1983).

7. Gelijkens, C. F., D. L. Smith, and J. A. McCloskey. "Capillary Gas Chromatography of Pyrimidines and Purines: N,O Peralkyl and Trifluoroacetyl-N,O-Alkyl Derivatives," J. Chromatogr. 255:291-299 (1981).

8. Stadler, J., "Quantitative Microdetermination of DNA by Two-Dimensional Electron Capture Gas Chromatography of the Thymine Constituent," Anal. Biochem. 86:477-489 (1978).

9. Ludewig, M., K. Dorffling, and W.A. Konig. "Electron-Capture Capillary Gas Chromatography and Mass Spectrometry of Trifluoroacetylated Cytokinins," J. Chromatogr. 243:93-98 (1982).

10. Sentissi, A., M. Joppich, K. O'Connell, A. Nazareth, and R. W. Giese. "Pentafluorophenylsulfonyl Chloride: A New Electrophoric Derivatizing Reagent with Application to Tyrosyl Peptide Determination by Gas Chromatography with Electron Capture Detection," Anal. Chem. 56:2512-2517 (1984).

11. Poole, C. F., and A. Zlatkis. "Derivatization Techniques for the Electron-Capture Detector," Anal. Chem. 52:1002A-1016A (1980).

12. Ehrsson, H., and B. Mellstrom. "Gas Chromatographic Determination of Amides After Perfluoroacylation," Acta Pharm. Suedica 9:107-114 (1972).

13. Nazareth, A., M. Joppich, S. Abdel-Baky, K. O'Connell, A. Sentissi, and R. W. Giese. "Electrophore-Labeling and Alkylation of Standards of Nucleic Acid Pyrimidine Bases for

Analysis by Gas Chromatography with Electron Capture Detection," J. Chromatogr. 314:201-210 (1984).

14. Mohamed, G. B., A. Nazareth, M. J. Hayes, R. W. Giese, and P. Vouros. "GC-MS Characteristics of Methylated Perfluoroacyl Derivatives of Cytosine and 5-Methyl Cytosine," J. Chromatogr. 314:211-217 (1984).

15. Fisher, D. H., J. Adams, and R. W. Giese. "Trace Derivatization of Cytosine with Pentafluorobenzoyl Chloride and Dimethylsulfate," J. Environmental Health Sciences 62:67-71 (1985).

16. Fisher, D., T. Trainor, P. Vouros, and R. W. Giese. In preparation.

17. Adams, J., M. David, and R. W. Giese. "Pentafluorobenzylation of O^4-Ethylthymidine and Analogs by Phase-Transfer Catalysis for Determination by Gas Chromatography with Electron Capture Detection," Anal. Chem., 58:345-348 (1986).

18. Rogers, E., and R. W. Giese. In preparation.

19. Everson, R. B., E. Randerath, R. M. Santella, R. C. Cefalo, T. A. Avitts, and K. Randerath. "Detection of Smoking-Related Covalent DNA Adducts in Human Placenta," Science 231:54-56 (1986).

20. Nehls, P., J. Adamkiewicz, and M. F. Rajewsky. "Immuno-Slot-Blot: A Highly Sensitive Immunoassay for the Quantitation of Carcinogen-modified Nucleosides in DNA," J. Cancer Res. Clin. Oncol. 108:23-29 (1984).

21. Giese, R. W. "Electrophoric Release Tags: Ultrasensitive Molecular Labels Providing Multiplicity," Trends in Anal. Chem. 2:166-168 (1983).

22. Joppich-Kuhn, R., M. Joppich, and R. W. Giese. "Release Tags: A New Class of Analytical Reagents," Clin. Chem. 28:1844-1847 (1982).

23. Abdel-Baky, S., N. Klempier, and R. W. Giese. "Diol and Olefin Electrophoric Release Tags," 191st ACS National Meeting, New York, April 13-18, 1986, Poster 36.

24. Whitehead, J. K. and H. G. Dean. "The Isotope Derivative Method in Biochemical Analysis" in Methods of Biochemical Analysis, Vol. 16, D. Glick, Ed. (New York: Interscience Pub., 1986), Chapter 1.

25. Grafstrom, H. C., A. Fornace, Jr., and C. C. Harris. "Repair of DNA Damage Caused by Formaldehyde in Human Cells," Cancer Research 44:4323-4327 (1984).

CHAPTER 5

QUANTIFICATION OF TISSUE DOSES OF CARCINOGENIC AROMATIC AMINES

Paul L. Skipper, Matthew S. Bryant, and Steven R. Tannenbaum

INTRODUCTION

Until recently, estimation of exposure to exogenous carcinogens has largely been a process of environmental sampling to
determine peak or average ambient concentrations combined with
a calculation of probable intake as a function of the route of
exposure. A potentially more accurate approach is the quantification of carcinogen-protein adducts formed with readily accessible proteins, such as hemoglobin, which could serve as in
vivo dosimeters [1]. This paper will discuss some of our recent efforts in developing this approach for the dosimetry of
one group of structurally related compounds, the aromatic
amines.

Quantification of protein-carcinogen adducts provides a
measure of the end product of a series of steps which begin
with intake of the carcinogen. They also include distribution
to different tissues, detoxification and excretion, metabolic
activation, and redistribution of reactive metabolites whenever
the target organ is different from the one in which metabolism
occurs. There is a potential for intra- as well as interindi-
vidual variability in all of these steps. In order for protein
adducts to be dosimeters in the sense that they measure applied
dose, it is essential that the overall relationship between
carcinogen intake and protein adduct formation is known and
reasonably well defined. On the other hand, we deliberately
choose to measure adducts formed by the same metabolites which
are believed to react with deoxyribonucleic acid (DNA), transforming normal cells into potentially malignant ones. Pre-
sumably, the number of such initiation reactions is more closely related to tumor induction than is carcinogen intake, so it
seems likely that the level of protein adducts will be better

than environmental concentration as an indicator of true tissue burden. It is expected that there will exist situations in which there is a preference for knowing one or the other of these two measures, exposure or tissue dose of reactive metabolites. Therefore, in addition to the development of detection methodology for measurement of protein adducts, we are also developing animal models to help define the relationship between carcinogen intake and resultant adduct levels. The present discussion will focus on quantification and detection methodology, as most of our work on animal models has been published elsewhere [2,3].

ENVIRONMENTAL OCCURRENCE OF AROMATIC AMINES

Workplace exposure to relatively high levels of 4-aminobiphenyl (4-ABP), 2-naphthylamine (2-NA), and benzidine led to their identification as human bladder carcinogens. As a result, large-scale production of these and related amines has been curtailed, and today it is difficult to find instances of similar exposures. At lower levels, however, much of the human population is still exposed to these carcinogens from at least two documented sources. There are also other, more speculative sources, the significance of which remains to be established.

Cigarette smoke is now known to contain many aromatic amines in addition to a wide variety of other toxic compounds [4]. These include aniline and methyl- and ethyl-substituted anilines, naphthylamines, and 2-, 3-, and 4-aminobiphenyl. These compounds are present in both the mainstream smoke and the sidestream smoke. (Mainstream smoke is the smoke which is drawn through the cigarette and inhaled by the smoker. Sidestream smoke arises from the burning tip of the cigarette and enters the atmosphere directly.) Monocyclic amines are typically observed at levels of 10-100 and 100-10,000 ng/cigarette, respectively. The bicyclic amines are present at lower concentrations: 1-5 ng/cigarette in mainstream smoke and up to 150 ng/cigarette in sidestream smoke.

A second route of exposure for much of the population is through the use of dyes in foods and cosmetics. For example, D & C red #33 is contaminated with trace amounts of aniline, 4-ABP, 4-aminoazobenzene, and benzidine [5]. The amounts of free amines present in these dyes are unlikely to be significant health hazards per se, but are of interest because their occurrence suggests the presence of so-called subsidiary dyes, which are composed of various amino residues not present in the primary dye. Indeed, tartrazine (FD & C yellow #6, a commonly used food dye) has been shown to contain up to 0.52% of the subsidiary dye derived from aniline [6]. In vivo studies [7] as well as cell culture studies [8] have demonstrated that many dyes can be converted metabolically to the amino residues from which they are constituted. Thus, if there is significant contamination of primary dyes with subsidiary dyes which contain carcinogenic residues such as 4-ABP, metabolic breakdown could lead to far greater exposures than suggested by the amount of contaminating free amine.

METABOLIC ACTIVATION OF AROMATIC AMINES

Some of the early steps in the hepatic processing of aromatic amines are illustrated in Figure 1. One of the first important reactions which can occur is conjugation of the amines to form more water-soluble derivatives such as sulfamic acids or

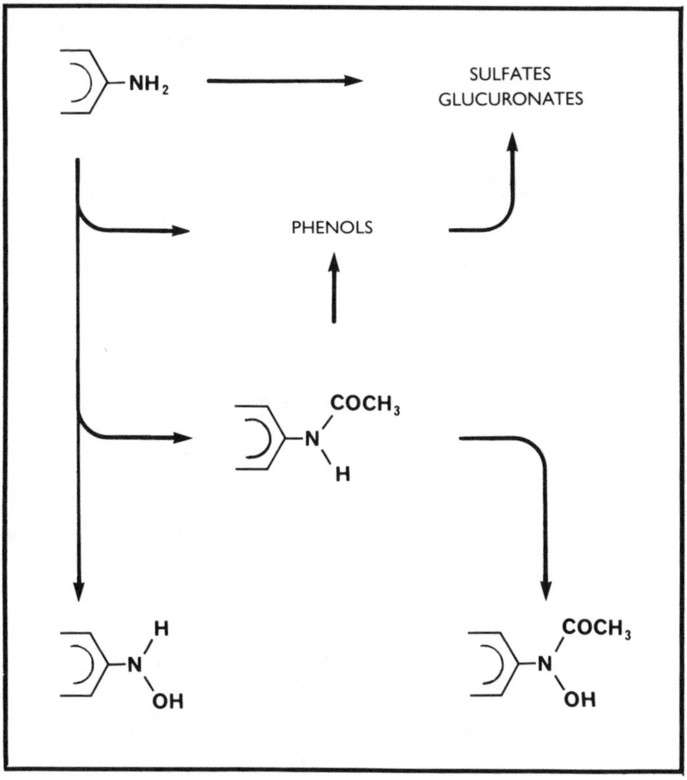

Figure 1.   Principal initial metabolism of aromatic amines. The products of N-hydroxylation may be toxic directly or undergo further metabolism to become toxic. Other products are generally excreted without reacting with cellular targets.

glucuronides which are excreted in urine or bile. C-hydroxyl-
ation to form phenols and conjugation of the phenols contribute
in a major way to overall detoxification and removal. Acetyla-
tion, however, continues the process in the direction of toxicity. It is probably a major determinant of organ specificity
in that non-acetylated metabolites are implicated in urinary
bladder carcinogenesis, whereas acetylated metabolites appear
to target other organs, such as the liver. Both amines and
acetamides share a common toxification reaction, N-hydroxylation. In some cases the resultant hydroxylamine or hydroxamic
acid reacts directly with cellular targets, and in other cases,
esterification or acyl transfer is necessary before reaction
occurs. In any event, the production of one of these two in-
termediates is probably obligatory for the ultimate formation
of any macromolecular adducts, whether they are formed with DNA
or with protein. For a comprehensive review of the metabolism
of aromatic amines, see, for example, Garner et al. [9].

With many aromatic amines, the major blood protein adduct
is formed from the N-hydroxylamine [10]. This adduct is a
sulfinic acid amide of the cysteine residues in hemoglobin.
Other adducts are also formed, with hemoglobin as well as with
other proteins, but not in comparable amounts. The hydroxamic
acids also produce adducts, but the present discussion will be
confined to the sulfinamide adducts, for which we now have a
method of detection sufficiently sensitive that for some
amines, even environmental levels can be quantified.

The cysteine sulfinamide adduct with hemoglobin is not produced by direct reaction of an aromatic hydroxylamine, but requires the intermediate production of the corresponding aromatic nitroso compound. This intermediate is the result of a
heme-mediated oxidation in which the heme iron is also oxidized, yielding methemoglobin. The immediate reaction product
of the nitroso arene with cysteine is the result of nucleophilic addition of sulfur at the nitrogen atom. This unstable
N-hydroxy sulfenamide rearranges to the more stable sulfinamide. The reaction sequence has been demonstrated in vitro
with thioglycerol and nitrosobenzene [11] and it is assumed
that the same occurs in vivo in hemoglobin.

Considerable circumstantial evidence has been accumulated
indicating that a specific cysteine residue, 93 in the beta
chain, is responsible for binding aromatic amines [12]. We
have been able to unequivocally demonstrate that this is true
for 4-aminobiphenyl by X-ray crystallography (unpublished results, this laboratory). The biphenyl residue was shown to oc-
cupy a space not existing in the native hemoglobin. That the
protein adopts a non-native configuration suggests strong non-
bonding interactions in addition to the covalent bond, and may
define the range of amines which can bind to hemoglobin as
cysteine sulfinamides, according to certain steric requirements.

METHODS OF ANALYSIS FOR CYSTEINE SULFINAMIDES

Detection and quantification of aromatic amine cysteine sulfinamides are greatly simplified by the relative chemical instability of these adducts. The in vitro rates of sulfinamide hydrolysis are much greater than the rates of peptide bond hydrolysis. The result of hydrolysis is to release the bound carcinogen as the free amine. The protein structure is preserved and the amines can be separated from the intact protein by organic solvent extraction. The analytical problem is thus reduced to that of detecting and quantifying free aromatic amines, separated from all the amino acid components of the original sample as well as practically all the incidental hydrophilic materials.

The sample workup consists of isolation of hemoglobin by standard methods, addition of appropriate internal standards, hydroxide-catalyzed hydrolysis, extraction, and an acid/base partition to remove neutral and acidic substances from the original extract. This is followed by derivatization with a perfluorinated acylating reagent and separation and quantification by capillary gas chromatography. The internal standards can be either the fully deuterated analog of the amine of interest or a monofluoro analog.

We have evaluated several derivatizing reagents, different types of capillary columns, and different types of detection. Of the derivatizing reagents, pentafluorobenzoyl chloride, although its derivatives do give the greatest response, is the worst because of the number of side products and impurities. Pentafluoropropionic and heptafluorobutyric anhydride are about equally good. The type of capillary column seems not to be very important, provided that it is a bonded phase type; the non-bonded phases have too much bleed. Polar and non-polar stationary phases both produce good peak shapes and give good but different resolutions. The choice may depend on particular circumstances. For instance, we use 4'-F-4-ABP as an internal standard for 4-ABP and this cannot be resolved from 4-ABP on a methyl phenyl silicone phase, but does separate well on a Carbowax™ phase.

Initially it was hoped that electron capture (EC) detection would be suitable, and some results using this technique have been reported [2,13]. The EC detector is relatively inexpensive and, in fact, is the most sensitive available for good electrophores. However, the sensitivity, which can be as good as 10 femtograms per injection, cannot be utilized fully because the lack of selectivity results in a too complex chromatogram. In practice, we found that the operational limit of detection, in relatively clean regions of the chromatogram, such as where 4-ABP elutes, is on the order of 100 pg/g hemoglobin [2].

The use of a mass spectrometer for detection solves the problem of insufficient selectivity, and with careful tuning, can provide more than adequate sensitivity if it is operated in a selected ion mode. All the common modes of ionization available for gas chromatography/mass spectrometry (GC-MS) can be used for detecting the perfluoroacyl aromatic amines. Electron impact ionization leads to the greatest number of fragment ions and is therefore inherently less selective than the others. Chemical ionization typically produces only one major and a small number of minor ions. In the positive chemical ionization mode, the most abundant ion produced by the PFP or HFB amines is M+H, in common with most other types of compounds. With negative chemical ionization, the most abundant ion is the M-20, corresponding to loss of hydrogen fluoride. Thus, although it is typically not possible to obtain full spectra from the amines at the levels at which they occur in human blood specimens, by switching ionization modes and detecting the characteristic major ions in each mode, it is possible to confirm the identity of the amine in question.

The method of choice is negative chemical ionization. It provides, at least in our hands, somewhat greater sensitivity than the other two. More importantly, whereas almost all substances will respond to electron impact or positive chemical ionization, far fewer respond well to negative chemical ionization. Figure 2 is a chromatogram obtained in the analysis of a human blood specimen for 4-ABP and illustrates the selectivity typically attainable with this method.

RECENT RESULTS

As indicated earlier, cigarette smoking is probably an important cause of human exposure to aromatic amines. Consequently, we have been interested in comparing smokers to non-smokers for the levels of hemoglobin-bound aromatic amines. Volunteers who will donate blood specimens and answer smoking history questionnaires are being recruited from several populations. In the first studies, we have examined the blood specimens for hemoglobin-bound 4-ABP. A total of 37 individuals was studied, consisting of 18 non-smokers and 19 smokers. The first group displayed an average adduct level of 32 ng/g hemoglobin (S.D. = 13). The group of smokers averaged 154 ng/g hemoglobin (S.D. = 47). The difference between the two means is highly significant (p = .0001), suggesting that for 4-ABP, cigarette smoking is the major contributor of exposure.

In addition to comparing smokers and non-smokers, we have been able to obtain blood specimens at selected intervals from five individuals who have stopped smoking. The average adduct level declined from 102 + 23 ng/g hemoglobin to 26 + 11 ng/g within 8 weeks after the individuals stopped and did not decline further after 4 months. The final value is much the same as that observed for non-smokers.

Preliminary data from some of the samples in the studies just described indicated that other aromatic amines were also

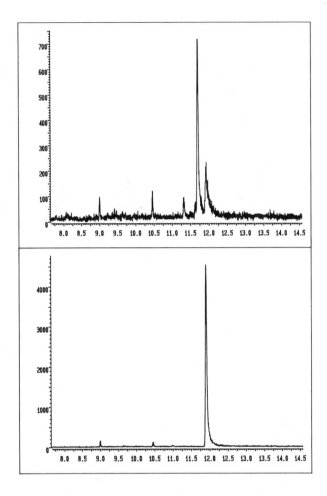

Figure 2. GC-MS analysis of 4-aminobiphenyl bound to hemoglobin in the blood of a smoker. Selected ion monitoring at m/z 295 (upper trace, 4-ABP) and m/z 313 (lower trace, 4'-fluoro-4-ABP, internal standard) was used for detection. The retention times, in minutes, were: 4-ABP, 11.68 and 4-fluoro-4-ABP, 11.91. The peak observed at 11.91 in the m/z 295 trace was produced by a minor M-38 fragment of the internal standard but cannot be used for quantification because the relative abundance is not constant. The adduct level in this sample was determined to be 108 pg/g hemoglobin.

present in the blood as hemoglobin adducts. These include 2-naphthylamine, aniline, and all three toluidines. It is premature to give quantitative assessments of levels because appropriate internal standards were not added to the samples at the time of workup. However, in future studies, internal standards will be included for those amines for which there is evidence of their presence.

Of perhaps equal interest are the negative findings that are accumulating. For instance, we are unable to detect ethyl anilines, aminoanthracene, or aminophenanthrene by the present method. This is true for blood specimens obtained from both smokers and non-smokers. There are, of course, numerous explanations for these findings, including relatively lower exposure to these amines, an inability of the reactive metabolite to bind to hemoglobin, instability of the amine after hydrolysis, or different metabolic processing. The last of these is discussed in more detail below.

DISCUSSION

It has been suggested [14] that the mechanism whereby aromatic amines such as 4-ABP and 2-NA induce cancer of the urinary bladder includes the following pharmacological steps. The amine is oxidized in the liver by one or more of the cytochrome P450 isozymes to the hydroxylamine, which is conjugated with glucuronic acid to form an N-glucuronide. The N-glucuronide, being stable at pH 7.4, serves to transport the reactive hydroxylamine to the bladder via the urine. Within the bladder lumen, where the pH is considerably lower than in blood, the N-glucuronide undergoes an acid catalyzed hydrolysis to liberate the hydroxylamine, which diffuses into and through the bladder epithelium into the blood. Within the epithelial cells it reacts with DNA, initiating tumorogenesis, and in the blood it reacts with hemoglobin.

If this explanation is correct, and there is considerable experimental evidence to support it, then the level of hemoglobin-bound aromatic amine should closely reflect the extent of exposure of the bladder epithelium to reactive metabolites. If there are no other significant pathways for entry of aromatic hydroxylamines into the blood, then hemoglobin sulfinamides would be expected to be observed primarily for those amines which exert a carcinogenic effect on the bladder and only secondarily for those amines which have other target organs.

ACKNOWLEDGMENTS

This work was supported by the National Institutes of Environmental Health Sciences Grant No. P01-ES00597, National Institutes of Health Training Grant No. 2T32 ES07020, and the National Cancer Society Grant No. SIG-10-I. The work described in this chapter was not funded by EPA and no official endorsement should be inferred.

REFERENCES

1. Ehrenberg, L., K. D. Hiesche, S. Osterman-Golkar, and I. Wennberg. "Evaluation of Genetic Risks of Alkylating Agents: Tissue Doses in the Mouse from Air Contaminated with Ethylene Oxide," Mutat. Res. 24:83-103 (1974).

2. Green, L. C., P. L. Skipper, R. J. Turesky, M. S. Bryant, S. R. Tannenbaum, and F.F. Kadlubar. "In Vivo Dosimetry of 4-Aminobiphenyl in Rats via a Cysteine Adduct in Hemoglobin," Cancer Res. 44:4254-4259 (1984).

3. Skipper, P. L., L. C. Green, M. S. Bryant, S. R. Tannenbaum, and F.F. Kadlubar. "Monitoring Exposure to 4-Aminobiphenyl via Blood Protein Adducts," in Monitoring Human Exposure to Carcinogenic and Mutagenic Agents (IARC Scientific Publications No. 59), A. Berlin, M. Draper, K. Hemminki & H. Vainio, Eds. (Lyon: International Agency for Research on Cancer, 1984), pp. 143-150.

4. Patrianakos, C., and D. Hoffmann. "Chemical Studies on Tobacco Smoke. LXIV. On the Analysis of Aromatic Amines in Cigarette Smoke," J. Anal. Toxicol. 3:150-154 (1979).

5. Bailey, J. E., Jr. "Determination of Unsulfonated Aromatic Amines in D&C Red No. 33 by the Diazotization and Coupling Procedure Followed by Reversed-Phase Liquid Chromatographic Analysis," Anal. Chem. 57:189-196 (1985).

6. Bailey, J. E., Jr. "Determination of the Lower Sulfonated Subsidary Colors in FD & C Yellow No. 6 by Reversed-Phase High-Performance Liquid Chromatography," J. Chromatogr. 347:163-172 (1985).

7. Walker, R. "The Metabolism of Azo Compounds: A Review of the Literature," Food Cosmet. Toxicol. 8:659-676 (1970).

8. Manning, B. W., C. E. Cerniglia, and T. W. Federle. "Metabolism of the Benzidine-Based Azo Dye Direct Black 38 by Human Intestinal Microbiota," Appl. Environ. Microbiol. 50:10-15 (1985).

9. Garner, R. C., C. N. Martin, and D. B. Clayson. "Carcin-
    ogenic Aromatic Amines and Related Compounds," in Chemical
    Carcinogens, Second Edition, C. E. Searle, Ed. (Washing-
    ton, D.C.: American Chemical Society, 1984), pp. 175-276.

10. Neumann, H.-G. "Analysis of Hemoglobin as a Dose Monitor
    for Alkylating and Arylating Agents," Arch. Toxicol.
    56:1-6 (1984).

11. Klehr, H., P. Eyer, and W. Schafer. "On the Mechanism of
    Reactions of Nitrosoarenes with Thiols," Biol. Chem.
    Hoppe-Seyler 366:755-760 (1985).

12. Kiese, M., and K. Taeger. "The Fate of Phenylhydroxyl-
    amine in Human Red Cells," Naunyn-Schmiedeberg's Arch.
    Pharmacol. 292:59-66 (1976).

13. Skipper, P. L., M. S. Bryant, S. R. Tannenbaum, and J. D.
    Groopman. "Analytical Methods for Assessing Exposure to
    4-Aminobiphenyl Based on Protein Adduct Formation" J.
    Occup. Med. In press.

14. Kadlubar, F. F., J. A. Miller, and E. C. Miller. "Hepatic
    Microsomal N-Glucuronidation and Nucleic Acid Binding of
    N-Hydroxy Arylamines in Relation to Urinary Bladder
    Carcinogenesis," Cancer Res. 37:804-814 (1977).

CHAPTER 6

THE FEASIBILITY OF CONDUCTING EPIDEMIOLOGIC STUDIES OF
POPULATIONS RESIDING NEAR HAZARDOUS WASTE DISPOSAL SITES

Gary M. Marsh and Richard J. Caplan

INTRODUCTION

The potential for hazardous wastes to cause health damage
to exposed human populations requires epidemiologic investigations to assess relationships between toxic exposure and possible health consequences, clinical or subclinical. Unfortunately, the classical application of epidemiology is made difficult under a myriad of methodologic complications and uncertainties related to both exposure and health outcome assessment. A particularly problematic feature of all health effects
evaluations at hazardous waste sites is the sheer diversity in
which toxic wastes and human exposures can be involved. Such
diversity not only prohibits the development of a unified analytic approach to exposure and health outcome assessment but
also prevents the generalization of statistical inferences
drawn about a specific waste site exposed population.
Regardless of the study design or diversity of the underlying setting, however, health effects evaluations of persons
exposed to chemical dumps consist of four fundamental phases:
documentation of the nature and extent of exposure, definition
and characterization of exposed and unexposed populations,
diagnosis and measurement of disease and dysfunction in the exposed population, and determination of the relationship between
exposure and disease [1,2]. This paper focuses on the general
epidemiologic considerations associated with the fourth phase
and proposes and evaluates specific classical and nonclassical
methodologic approaches to health evaluations. Primary consideration is given to the health effects of continuous low dose
chemical exposures of a noninfectious nature that originate

from existing common sources (dump sites, lagoons, landfills, etc.) as opposed to temporary, intermittent, or episodic exposures resulting from accidental spills or discharges.

DETERMINATION OF EXPOSURE-HEALTH OUTCOME RELATIONSHIPS

The ultimate objective in epidemiologic studies of persons exposed to hazardous waste site materials is to associate particular exposures with potential biologic effects and thus identify cause-effect relationships. Such associations are considerably strengthened if dose-response relationships can be found, that is, if increasing levels of exposure are associated with increasing frequency of the biologic effect. The achievement of this objective is made difficult, however, not only by the limitations which are inherent in all observational studies of human populations, but also by the number of particularly complex real-life situations which uniquely characterize waste site studies. In this context, the fullest exploration of human observational studies is often greatly restricted by concern for confidentiality on the part of exposed and affected persons, for parsimony by health authorities, for safeguards on the part of industry, and for political considerations on the part of government agencies. More specifically, epidemiologic studies of populations exposed to toxic waste site materials are likely to be limited by the following technical and human problems [3,4]:

o Populations living in the vicinity of a hazardous waste site are usually small, thus limiting both the range of outcomes and the size of the effects that can be studied.
o Persons living in any given area are usually heterogeneous, either with respect to characteristics that can influence many health outcomes independently of exposure (age, race, socioeconomic status, occupation, smoking, alcohol consumption, etc.), or with respect to the type, level, duration, or timing of exposure. Moreover, there is in-and-out migration and geographic mobility within areas.
o Actual population exposures are generally poorly defined.
o Many of the health endpoints of interest are either rare (such as specific malformations), are associated with long or variable latency periods (such as cancer), or are unlikely to have been routinely recorded prior to the investigation (such as spontaneous abortions). In addition, the instruments used to measure health outcomes (e.g., questionnaires) are generally very insensitive.
o Publicity related to the episode under study may produce or accentuate reporting bias.
o The conduct of waste site studies is made difficult due to the presence of a highly charged atmosphere of anger

and fear which often accompanies suspicion of adverse health effects. Moreover, in some cases otherwise unwarranted studies are mounted in reaction to existing public concern in an area.

The following section describes how the aforementioned methodologic problems can affect the statistical aspects of any well-designed study of toxic exposures, in particular, statistical power, bias, and interaction. How these methodologic limitations affect the choice, conduct, and statistical aspects of specific epidemiologic study designs is discussed in the next major section to follow.

STATISTICAL POWER

In the context of hazardous waste site studies, statistical power can be defined as the probability that an adverse health effect of a specific size will be detected when it is present in the target population from which the sample was drawn. Power is an extremely important consideration, since it helps to determine study design and provides an objective basis from which to meaningfully interpret study results. Statistical power is a function of the following study parameters:

o   The size of the study and control groups. In general, power increases as the size of the population under study increases.
o   The variability of the health outcome under study. For discrete events this will depend on the usual or expected rate of the event in the control population. In general, power is inversely related to the variability of the health outcome in the target population.
o   The predetermined statistical significance level or Type I error that will be accepted as confirmation of an association between exposure and health outcome. This assumes a specific probability model, of which more than one may be feasible. With all other parameters fixed, power is directly related to the significance level.
o   The magnitude of the expected association between exposure and outcome. With all other parameters fixed, power is directly related to this magnitude.
o   The design of the study and statistical techniques used for analysis.

There are several special design and analytic techniques that may be used to enhance power. These include refining the history of exposure to avoid misclassification bias; refining the response variable to conform with an anticipated biologically coherent health outcome; increasing the study population size via intensified case finding; forming composite exposure or outcome variables; use of continuous rather than discrete

health outcome variables; use of repeated measures on each
study member; stratification or matching; and clustering techniques [3,4].

The interrelationships of the primary study parameters that
determine statistical power are illustrated in Figures 1 and
2. Figure 1, which pertains to cohort or cross-sectional studies, shows the relationships between health outcome frequency,
sample size (in study and control group), and the magnitude of
the effect that can be demonstrated at the two-tailed 5% significance level with a power of 80% [5]. In general, Figure 1
shows that for a given sample size the power to discern modest
effects increases with increasing frequency of the event under
study, or for a given level of frequency the ability to detect
a given effect size increases with increasing sample size.

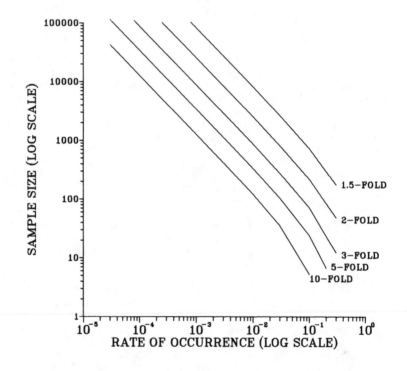

Figure 1.   Rate of occurrence vs. required sample size to de-
             tect an N-fold increase in rate. Alpha = .05 (2-
             tailed), power = .80.

In a similar fashion, Figure 2 shows, for unmatched case-
control studies, the relationship between the proportion of
controls exposed, sample size (in case and control group), and

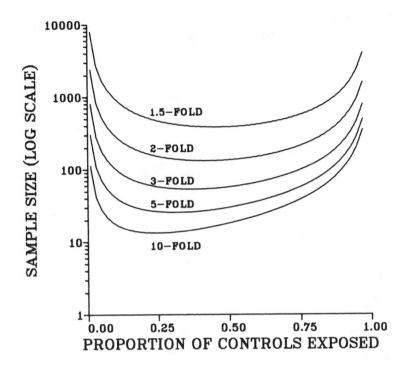

Figure 2. Proportion of controls exposed vs. required sample size to detect an N-fold relative risk. Alpha = .05 (2-tailed), power = 80.

the minimum relative risk that can be detected at the two-tailed 5% significance level with a power of 80% [6]. In general, Figure 2 shows that for a given case and control group size, the power to discern modest effects is maximized when the proportion of exposed controls is around 0.25 to 0.50. Conversely, for a given proportion of exposed controls, the ability to detect a given effect size increases with increasing sample size. In order to place Figures 1 and 2 into proper perspective, Table 1 provides the frequencies of occurrence of selected health outcomes that might be examined in a waste site study.

Although the above illustrative samples demonstrate that large (i.e., 10-fold) increases in background disease frequency could be detected at acceptable error levels in cohort or case-control studies with quite manageable sample sizes, it is unlikely that the relatively low levels of toxic chemical exposures that prevail in most waste site situations would produce excesses in disease of this magnitude.

Table 1. General Background Frequencies and Units of Analysis
of Selected Health Outcomes.

Health Outcome	Frequency	Unit of Analysis
Reproductive Effects[a]		
Azoospermia	1×10^{-2}	Males
Birthweight < 2500 g	7×10^{-2}	Livebirths
Spontaneous abortion 8-28 weeks of gestation	$1-2 \times 10^{-1}$	Pregnancies
Chromosomal anomaly among spontaneously aborted conceptions	$3-4 \times 10^{-1}$	Spontaneous abortion
Birth defects	$2-3 \times 10^{-2}$	Livebirths
Neural tube defects	$1 \times 10^{-4}-1 \times 10^{-2}$	Livebirths and stillbirths
Cancer Incidence[b]		
All sites	3.2×10^{-3}	Individuals
Stomach	9.8×10^{-5}	Individuals
Colon	3.3×10^{-4}	Individuals
Lung and bronchus	4.5×10^{-4}	Individuals
Bladder	1.5×10^{-4}	Individuals
Kidney	6.4×10^{-5}	Individuals
Lymphomas	1.2×10^{-4}	Individuals
Leukemias	9.5×10^{-5}	Individuals
Mortality[c]		
All causes	9.5×10^{-3}	Individuals
All cancer sites	1.6×10^{-3}	Individuals
Cirrhosis of liver	1.5×10^{-4}	Individuals
Congenital anomalies	8.4×10^{-5}	Individuals

[a]From Reference 4.
[b]Average annual age adjusted (1970) incidence rates, 1973-76,
all SEER sites, Reference 7.
[c]1970 U.S. mortality rates.

BIAS

Since the ultimate aim of any study is to describe an
exposure-outcome relationship that is unlikely to be explained
by extraneous differences between the two study groups, it is
imperative that two sources of variation be controlled: variation in the characteristics of the study groups that relate to
the a priori chance of exposure or to outcome and variation in
the quality of data collected for the two study groups [3].
The inability to control for these sources of variation can

lead to a biased estimate of the exposure-outcome relationship. These and other sources of bias resulting from methodological features of study design and analysis are further discussed and classified in the catalog of biases provided by
Sackett [8].

INTERACTION

Generally, discussions of determining exposure-outcome relationships devote little attention to the problems of multiple
concurrent exposures. As noted above, individuals living in
the vicinity of hazardous waste sites may be exposed via many
different routes to mixtures or combinations of several potentially toxic chemicals.
Several authors have considered the statistical-
epidemiologic issue of interaction as it applies to the combined effect of two or more exposures [9-12]. Although the
requisite analytic methods have been developed to assess interactions, their application to waste site studies may be limited. That is, the most useful and interpretable analysis of
interaction requires the application of multivariate statistical techniques, and the typically weak and incomplete data derived from waste site studies may not be amenable to these more
sophisticated modes of analysis. Also, several types of statistical models are available for assessing interaction, but
there is some dispute over which is the most appropriate
[10,11,13]. The relevancy of these statistical models to the
biology of waste site related illness will remain uncertain,
however, until more is known about how various chemical exposures produce illness or biologic effects [14].

SPECIFIC METHODOLOGIC APPROACHES TO HEALTH EFFECTS EVALUATIONS

The Level of the Investigation

Based primarily on practical considerations, health effects
investigations can be classified into three levels [4]. Level
I is based on existing, routine, and easily accessible exposure
and health outcome records. The investigation will usually be
conducted with speed and economy and will seldom involve case
examinations or special questionnaires. Level I studies will
lack power, since they will usually be limited to poorly defined measures of exposure. They may also be deficient by
being unable to adjust estimates of exposure-outcome relationships for the effects of potentially confounding factors.
Since they generally involve aggregate versus individual data

on exposure and health outcome, Level I investigations include the large class of ecologic studies. For example, Level I studies may draw on vital certificate data or special registries of tumors or malformations in order to examine birthweights, sex ratios, perinatal mortality, and cancer incidence or mortality.

Level II includes short-term purposeful epidemiologic studies, such as cross-sectional, case-control or short-term cohort, that require the collection of more precise, individual exposure and health outcome data as well as data on potentially confounding variables. In Level II studies, the statistical considerations of power, bias, and interaction discussed above are applicable to the choice of study design, enabling the researcher to make maximal use of small numbers, rare events, and uncertain information sources. Level II studies can entertain a wide range of endpoints and can include outcomes identified through medical records (spontaneous abortions, malformations, behavioral or psychological disorders), through interviews with study subjects (spontaneous abortions, sexual dysfunction, symptoms or signs of rashes, paralysis, eye irritation, etc.), or through biological studies of study subjects (biochemical, immunologic and chromosomal assessments, and nerve conduction velocities) [4].

Level III involves well-planned, long-term investigations such as prospective studies of exposed and unexposed residential cohorts. Since this design is well suited for diseases with long latency periods, it has been considered mainly for the purpose of discovering environmental carcinogens [4]. Level III studies are greatly facilitated by the existence of centralized and accessible registries of births, deaths, and diseases, such as the National Death Index [15] recently created by the National Center for Health Statistics.

An example of a current waste site related investigation that encompasses all three levels of study is provided by the Centers for Disease Control's (CDC's) proposed study of PCB-exposed cohorts [16]. This study proposes a systematic approach to evaluate the degree of human exposure and the extent of health effects at Superfund sites associated with elevated levels of PCB:

o ecological assessment (Level I effort) to identify sites with PCB exposures;

o pilot exposure study (Level II effort) to document body burdens of PCB among the "most exposed" persons at each site;

o community survey (Level II effort) to identify cohorts of PCB exposed persons with little or zero levels of other toxic chemicals which would confound the health effects of PCB; and

o cohort study (Level III effort) to design and conduct registries of PCB exposed cohorts detected in the third stage in order to examine the long-term health effects of low-level PCB exposure.

Classical Epidemiological Study Designs

This section highlights the advantages of several classical study designs that may be utilized to evaluate health effects at hazardous waste sites. Basic epidemiologic designs consist of three types of studies (given the most emphasis in standard texts): cohort (follow-up), cross-sectional, case-control (independent and nested) [17-19]. The objectives of the first two types may be descriptive or etiologic, whereas the objectives of case-control studies are traditionally etiologic. In addition, this section considers some potentially useful "incomplete" designs or studies in which information is missing on one or more relevant factors. Finally, the utility of population registries for long-term follow-up studies is discussed.

Basic Designs

Cohort Studies. In this design, information about exposure status is known for all subjects at the beginning of the follow-up period. Both exposed and unexposed study members are followed for a given period of time for comparison of risks of developing a health outcome of interest. The health outcome may be cases of disease (incidence) or death (mortality), identified through reexaminations or population surveillance. In cohort studies, the unexposed group may be defined as having no exposure to the agent under study (e.g., comparison of persons in an exposed community with persons in an unexposed community) or as exposed at lower doses than the exposed group (e.g., persons residing at varying distances from the site of environmental contamination). Since it includes exposure data which is measured before or during the observation of health outcomes, the prospective design is generally preferred over other designs for making causal inferences. While a cohort study may be conducted prospectively or retrospectively, the latter approach is usually more cost- and time-efficient for studying rare diseases or diseases associated with long latent periods. The retrospective or historical-prospective design depends strongly on the availability of previous exposure information on a well- defined population that has been followed for detection of new cases or deaths [17].

The chief advantages of the general cohort design are that the relative and attributable risks are directly estimable as measures of association, incidence as well as mortality can be measured, and that a wide variety of health outcomes can be examined within a single study. The major weakness of the cohort study design is that it is statistically and practically inefficient for studying rare diseases.

There are currently several attempts planned or underway to utilize this design in waste site exposed environments where

the population at risk can be identified and followed and surveillance of specific diseases can be done. One example involving the historical-prospective design is the proposed mortality study of Talbott and Radford [20] of a community exposed
since 1910 to low-level radon and gamma radiation. This study
will attempt to evaluate the total and cause-specific mortality
experience of a community cohort previously identified by
Talbott et al. [21] via the cross-sectional approach as having
a higher rate of radiation-related thyroid diseases as compared
to residents of a nearby unexposed community. Their proposed
mortality study includes an exposed cohort of 6000 persons living as of the year 1938 within a one-mile radius of a uranium
waste site (near Canonsburg, Pennsylvania) and a control cohort
of 6000 persons living at the same time in a nearby unexposed
community (Bridgeville, Pennsylvania). As another example, the
Pennsylvania Department of Health developed a protocol to systematically investigate the health status of former employees
and selected residents of Lock Haven, Pennsylvania, who may
have experienced hazardous exposures associated with the Drake
Chemical Superfund site [22]. This protocol included both a
conventional retrospective occupational cohort study as well as
a current and retrospective community-based cohort study performed by Logue et al. [23] of households in the immediate vicinity of the Drake site. In addition to the cohort study, the
Drake site protocol included a cancer mortality and congenital
malformation incidence review, a health questionnaire survey,
and a bladder cancer screening component.

Cross-Sectional Studies. In this design (either nondirectional
or backward) a study population is selected from a single target population. This design involves the prevalence of health
outcomes, rather than the incidence, and usually involves random sampling of the target population. The backward design
begins with the classification of disease or dysfunction (e.g.,
case versus noncase) and proceeds by obtaining, though interview or examination, information about individual histories of
the study factor (i.e., previous exposures, events, or characteristics). In the nondirectional design, both the study factor and the disease are observed simultaneously, so that
neither variable may be uniquely identified as occurring
first. The utility of cross-sectional studies for describing
the frequencies of health outcomes or other characteristics and
for making causal inferences is severely limited if random
(probability) sampling is not incorporated into the design.
    Since the cross-sectional design does not involve a follow-
up period, it is often used to generate new etiologic hypotheses regarding study factors and/or health outcomes. Cross-
sectional studies are particularly useful for studying conditions that are quantitatively measured and that can vary over
time (e.g., blood pressure) or relatively frequent diseases
that have long duration (e.g., chronic bronchitis). They are
not appropriate for studying rare diseases or diseases with

short duration. Because the results of cross-sectional studies
are usually largely derived from interview data, they are especially prone to the methodologic limitations associated with
nonresponse, nonspecificity of health outcome and/or exposure,
recall bias, and the necessity of working in a highly charged
emotional atmosphere.

Despite its many limitations, the cross-sectional design
incorporating questionnaires has been utilized in many of the
waste-site-related health studies published to date.

Case-Control Studies. This study type involves a backward or
nondirectional design that compares a group of cases with a
specific disease and one or more groups of noncases without the
disease (controls) with respect to a current or previous study
factor level (exposure). A fundamental difference between this
study design and the cross-sectional is that the study groups
in the classical case-control design are selected from separate
populations of available cases and noncases, rather than from a
single target population. The control group may be derived
from a number of sources including hospitals, neighborhoods, or
the general population from which the cases were identified.
Since it is usually not possible at the outset of a study to
ascertain the comparability of cases and controls with respect
to potentially confounding variables, efforts are generally
made to control for confounding bias either through design
(matching cases to one or more controls on the basis of one or
more confounding characteristics) or through analysis (stratification by levels of one or more confounding characteristics).

The primary advantages of case-control studies over other
designs are that they are well suited to testing etiologic
hypotheses for specific rare diseases and that they allow for
the investigation of diseases with any latent period or duration of expression. The principal limitations of case-control
studies are the potentials for recall bias in exposure assessment and that only one health outcome of interest can be enter-
tained within a particular case-control study.

To date, the classical case-control approach has not been
widely applied to evaluate health effects of waste site or
other toxic environmental exposures. However, there have been
applications of a hybrid-case-control/cohort design, as discussed below.

Nested Case-Control Studies. This approach combines a few of
the major advantages of both cohort and case-control studies.
In this design, a single population is defined at the onset
without regard to exposure information, and is followed for a
given period for the detection of all incident cases or
deaths. The incident cases or deaths are then compared with a
group of controls sampled from the same population with respect
to previous or current exposure levels. The controls may be

sampled randomly from the population or they may be matched to
the incidence cases or deaths. The nested case-control design
is usually applied when an etiologic hypothesis emerges after
the beginning of follow-up or when limited resources preclude
the measurement of exposures on every subject in the study population.

The major advantage of the nested case-control design over
independent case-control designs is the assurance that cases
and controls are identified from the same well-defined population. Furthermore, since exposure information is obtained only
on a small fraction of the noncases in the study population,
this design, unlike the cohort design, is suitable for studying
rare diseases.

An appropriate situation for an ambidirectional study is
one in which it is possible to identify most new cases (or
deaths) of one or more rare diseases in a large population by
using existing information systems, such as employment or insurance records, a disease registry, or vital records. An example of such an application is the work of Lyon et al. [24]
who studied cancer clustering around a coke oven and uranium
tailing dump. In this study, the distribution of distances to
the point source of exposure (i.e., the exposure variable) for
cases of lung cancer in a two-county area between 1966 and 1975
was compared to the distribution for a control group of other
cancer cases that occurred in the same area and time period.
Both the cases and controls were drawn from the Utah Cancer
Registry.

Incomplete Designs

Incomplete designs, being Level I investigations, are frequently used when data are not readily available for conducting
another type of study. It is often relatively inexpensive or
convenient to utilize secondary data sources to test or generate etiologic hypotheses via these designs before considerable
time and resources are allocated to primary data collection.
This section considers two of several classes of incomplete designs for potential application to hazardous waste site health
effects evaluations. In addition, reference is made to several
secondary existing data sources that may be incorporated into
these designs.

Ecologic Studies. Broadly defined, these studies are empirical
or descriptive investigations involving the group as the unit
of analysis. Typically, the group is a geographically defined
area such as a state, county, or census tract. Ecologic analysis may involve incidence, prevalence, or mortality data, but
the latter is most common because of the widespread availability of such data. The primary analytic feature of an ecologic

study is the lack of information about the joint distribution of the study factor and the disease within each group (i.e., unit of analysis). Morgenstern [25] provides an excellent discussion of ecologic studies in a recent review article.

Ecologic studies are well suited as a preliminary or exploratory approach to evaluating health effects of waste site exposures. However, despite their practical advantages, causal inference about individual events from grouped data is limited by the heterogeneity usually found among groups (the so-called "ecologic fallacy") and the interrelationships which commonly exist among certain predictor variables (multicollinearity) [25].

The utility of ecologic analysis for evaluating health effects at hazardous waste sites depends heavily upon the availability of published summary data on exposure and/or health outcome that are specific for an appropriate unit of analysis. The National Priority List data bases and the centralized toxicological data banks, for example, are national-level data repositories that may provide useful summary data on potential exposure specific to geographic areas that contain toxic waste sites. On the other hand, the availability and accessibility of ecologic (or individual) data on health outcomes relevant to waste site studies vary according to geographic area and type of outcome. Also, the National Center for Health Statistics (NCHS) publishes summary vital statistics data collected through states on numerous topics. This is a particularly good source for determination of state, metropolitan, and national birth and death rates. Moreover, much of the NCHS data is available at the detailed individual record level on magnetic tapes, which can be purchased through the National Technical Information Service [26].

To overcome the lack of specificity inherent in published mortality rates (e.g., age-specific death rates at the county level are not published) several institutions including the University of Pittsburgh Graduate School of Public Health and the Johns Hopkins School of Hygiene and Public Health have linked the NCHS detailed mortality data with detailed U.S. census population data to develop computerized data retrieval/rate generating systems [27,28]. For example, the Mortality and Population Data Base System (MPDS) developed by Marsh et al. at the University of Pittsburgh can generate state, county, age, race, and sex-specific death rates for the years 1950-82 for any cause of death (cancer deaths only for 1950-62) specified by the appropriate four digit International Classification of Diseases (ICD) code.

The availability of ecologic data on morbidity, in particular cancer incidence data, is much more dependent upon the geographic area of study. For example, cancer incidence data developed through NCI's Surveillance, Epidemiology, and End Results (SEER) Program is available for certain years for only about 10% of the United States population which resides in the major Standard Metropolitan Statistical Areas (SMSA) [7]. Cancer incidence data for other geographic areas such as states, counties, or localities are available only for those states or subdivisions thereof that have developed tumor registry programs. Currently, tumor registries exist or are under development in about 12 states. Greenberg et al. [29] provide an extensive review of the measurement, sources, and uses of cancer incidence data in the United States.

Data systems are also available that integrate mortality and morbidity statistics with various sources of environmental data. The data included on two such systems, the Socio-Economic-Environmental Demographic Information System (SEEDIS) [30,31] and UPGRADE [32,33] are described in a recent review article by McCrea-Curnen and Schoenfeld [34].

Proportional Studies. These studies include observations on incident cases or deaths without information about the candidate population at risk of developing the health outcome(s). Due to the availability of mortality data, the proportional mortality design has been more widely applied, particularly in studies of occupational groups (for examples see references 35 and 36). The basic approach of the proportional study is to compare the proportion of total cases (or deaths) resulting from the disease of interest among different levels of exposure. From this approach, therefore, it is only possible to test the exposure-outcome relationship of primary interest if it can be assumed that there is no association between the exposure variable and the remaining (or comparison) diseases. Due to the limitations associated with using mortality and most morbidity data in waste site health effects evaluations, the utility of the proportional design is limited.

Establishment of Registries of Potentially Affected Persons

Although the establishment of registries of persons possibly exposed to toxic waste site materials is similar to the determination of the exposed and unexposed groups in a cohort study, there are two basic distinctions between the two approaches.

First, in the cohort study the exposure status of each person is known at the onset of the study. Exposure status of persons enrolled in a registry may or may not be known until subsequent examinations/interviews are conducted. Second, exposure and health outcome data collected on subjects in a cohort study may not be maintained or updated after completion of the study. The registry, on the other hand, can be considered as an open-ended cohort study, since it provides a general data base of exposed and unexposed persons that can be exploited in numerous ways to determine possible consequences of chemical or other exposures.

In general, therefore, with a registry it is not necessary to develop a specific study protocol until after the potentially exposed persons have been identified. This characteristic is most advantageous, since with the rapid mobility of the general population, the identification of persons possibly exposed to the hazard present at a waste site must be made as soon as possible following recognition of the problem. Thus, ideally, if a particular waste site is suspected of posing a hazard to human health, a registry of potentially affected people should be established as one of the first courses of action.

Unlike the site-specific registry described above, an exposure-specific registry is one which assembles persons from two or more locations on the basis of their common exposure to one hazardous material. While such registries are homogeneous with respect to exposure, they are more likely to be much less homogeneous with respect to other factors that might potentially confound an exposure-outcome relationship. An example of an exposure-based registry is the registry of PCB-exposed persons recently proposed by CDC [16].

The establishment of a registry can also provide crucial information needed by various state and federal agencies who may recognize the need for epidemiologic studies for research, or for determining who might require health care and long-term follow-up. It is also important to recognize that not all waste site situations are amenable to or even require the establishment of registries. In particular, the long-term usefulness of registries may be restricted by the inability to locate persons who have migrated out of the study area. Although there are centralized federal, state, and local sources that can be utilized for tracing individuals (e.g., Social Security Administration, state drivers license bureaus, etc.), they may require key record linkage elements (e.g., Social Security numbers) that may not be available from existing record sources. To date, the federal government and several state health departments have initiated the establishment of registries of persons possibly exposed to hazardous waste site materials.

Table 2 provides an outline of the aforementioned classical epidemiologic study designs according to the level of the investigation.

Alternative/Nonclassical Approaches

It has been shown throughout this paper that the application of classical epidemiological methods to evaluate health effects at hazardous waste sites is made difficult due to a wide variety of methodological limitations and particularly complex real-life situations. In view of this dilemma, it is

Table 2.   Outline of Classical Epidemiologic Study Designs by
Level of Investigation.

Level I (Based on Existing Exposure and Health Outcome Records)

A. Ecologic studies
B. Proportional studies

Level II (Short-term Designed Epidemiologic Studies)

A. Cohort (follow-up) studies
   1. Retrospective (historical-prospective) design
   2. Prospective design
B. Cross-sectional studies
   1. Backward design
   2. Nondirectional design
C. Case-control studies
   1. Backward design
   2. Nondirectional design
   3. Nested case-control studies

Level III (Long-Term Designed Epidemiologic Studies)

A. Cohort (follow-up)
B. Population-based registries
   1. Exposure-specific
   2. Site-specific

---

crucial that environmental epidemiologists begin both to develop methods to enhance the analytic capabilities of the classical approaches and to consider alternative methodologic approaches that will pave the way to a more complete understanding and perhaps an ultimate solution of waste site related
health problems.   In general, there are four categories of
"nonclassical" approaches that might be pursued:

1. to increase the inferential capabilities of existing
   statistical/epidemiologic methods by increasing analytic control over extraneous factors (e.g., multivariate methods) or by decreasing the dependency of the
   methods to underlying assumptions or requirements
   (e.g., development of nonparametric alternatives).
2. to explore familiar roles for epidemiology in nonenvironmental settings and to utilize these roles as par-
   adigms for possible roles in hazardous waste site settings.
3. to consider other nonepidemiologic methods that are
   used to assess analogous problems in nonenvironmental
   settings.

4. to employ the classical methods of epidemiology to study the health experience of occupational groups that are heavily exposed to substances commonly found in waste sites.

Although the statistical/epidemiologic literature abounds with activities and developments related to the first category of enhanced classical approaches, there is inevitably a prolonged lag before such refinements become a routine component in actual research problems. For example, in recent years there has been a considerable growth in the body of knowledge related to case-control methodology (e.g., linear logistic regression techniques, log linear modeling, and proportional hazards modeling); however, most of these newer methods require a level of mathematical and computer programming sophistication that impedes their rapid dissemination and application to real-life problems. Researchers in environmental epidemiology should make concerted efforts to regularly review the relevant literature in order to expeditiously exploit to the fullest extent possible any new methodologies that could be brought to bear on hazardous waste site epidemiology.

The second general category of alternative approaches was discussed by Neutra at the 1981 Rockefeller Symposium [37]. Arguing by analogy of the function of epidemiology in the infectious disease field, Neutra relates a paradigmatic model for the natural history of infectious disease to an analogous model for the natural history of chemically induced illness.

At the same symposium, Selikoff also endorsed the need for this second category of alternatives by advocating the development and application of approaches such as seroepidemiolgy, biochemical epidemiology, and epidemiological immunotoxicology to health effects evalautions at waste sites [38].

Compared to the first two categories, the third category of alternative approaches (i.e., nonepidemiologic methods) to waste site related health evaluations has probably received the least attention in the scientific literature. Neutra also alludes to this category by suggesting, for example, that data on subjective symptomatology (which is prevalent in health surveys of waste site exposed communities and is often viewed as psychosomatic or hypochondriacal) be subjected to nonclassical epidemiologic techniques, such as numerical taxonomy [39] (a type of cluster analysis) and discrimination function analysis or principal components analysis [40] in order to assess whether patterns of simultaneous symptoms differ for exposed and unexposed groups. In an analogous fashion, the techniques of numerical taxonomy and discriminate analysis have been used quite successfully in the epidemiology of colitis [41] and other poorly understood syndromes to assess whether reported complaints constitute any recognizable syndromes.

The fourth category of alternative approaches is fundamentally different from the others, since it does not directly involve the waste site exposed community, but rather a surrogate target population that has received similar, albeit more intense exposures. Occupational groups, in general, are especially amenable to epidemiologic inquiry since historical records are often available that permit the construction of well-defined cohorts which can be studied for diseases that are relatively rare and/or are associated with long latency periods. Moreover, since historical exposures received by cohort members are often documented or can be adequately extrapolated from current measurements, the health outcomes among occupational cohorts can often be related to type, duration, and intensity of exposure(s).

Thus, the study of an occupational group that has received relatively heavy exposures to a waste site related agent of interest enables a determination to be made of possible biological endpoints and the examination of dose-response relationships, so that at least an effort at extrapolation to the community population can be made.

The utility of this fourth approach has been recognized by many investigators, for example, the Pennsylvania Department of Health, which included both an occupational and community cohort in its health study protocol for the Lock Haven, Pennsylvania, Superfund site [22,23]. Perhaps the ideal occupational groups for study, however, are those whose work involves direct exposure to waste site materials, such as equipment operators who clean or maintain waste sites or workers aboard incinerator ships at sea [42].

The continued development of all categories of effective nonclassical approaches will require, at a minimum, increased communication and collaboration among researchers from a variety of allied professions, including epidemiology, medicine, biostatistics, mathematics, engineering, and toxicology. In this spirit, at least four recently published meetings [43-46] have hopefully provided the groundwork for future collaborative efforts in the hazardous waste site area.

ACKNOWLEDGMENT

This research has been funded in part by the U.S. Environmental Protection Agency under assistance agreement number CR811173-01 to the Center for Environmental Epidemiology at the University of Pittsburgh.

REFERENCES

1. Landrigan, P. J. "Epidemiologic Approaches to Persons with Exposures to Waste Chemicals," *Environ. Hlth. Persp.* 48:93-97 (1983).

2. Heath, Jr., C. W. "Field Epidemiologic Studies of Populations Exposed to Waste Dumps," Environ. Hlth. Persp. 48:3-7 (1983).

3. Stein, Z., M. Hatch, J. Kline, P. Shrout, et al. "Epidemiologic Considerations in Assessing Health Effects at Toxic Waste Sites," in Assessment of Health Effects at Chemical Disposal Sites, Proceedings of a Symposium at Rockefeller University, June 1-2, 1981. (New York: Rockefeller University, 1981).

4. Bloom, A. D. (Ed.) "Report of Panel II. Guidelines for Reproductive Studies in Exposed Human Populations," in Guidelines for Studies of Human Populations Exposed to Mutagen and Reproductive Hazards. (White Plains, NY: March of Dimes Birth Defects Foundation, 1981.)

5. Fleiss, J. Statistical Methods for Rates and Proportions, 2nd. (New York: John Wiley & Sons, 1981).

6. Schlesselman, J. J. Case-Control Studies. Design, Conduct, Analysis. (New York: Oxford University Press, 1982).

7. Young, Jr., J. L., A. J. Asive, and E. S. Pollack, Eds. "SEER Program: Cancer Incidence and Mortality in the U.S., 1973-76," Biometry Branch, Division of Cancer Cause and Prevention, National Cancer Institute. DHEW Publ. No. 78-1837 (1978).

8. Sackett, D. L. "Bias in Analytic Research," J. Chronic Dis. 32:51-63 (1979).

9. Koopman, J. S. "Causal Models and Sources of Interaction," Am. J. Epid. 106:439-444 (1977).

10. Kupper, L. L. and M. D. Hogan. "Interaction in Epidemiologic Studies," Am. J. Epid. 108:447-453 (1978).

11. Rothman, K. J. "Synergy and Antagonism in Cause Effect Relationships," Am. J. Epid. 99:385-388 (1974).

12. Rothman, K. J. "Occam's Razor Pares the Choice Among Statistical Models," Am. J. Epid. 108:347-349 (1978).

13. Walter, S. C. and T. R. Holford. "Additive, Multiplicative, and Other Models for Disease Risks," Am. J. Epid. 208:314-346 (1978).

14. Kimbrough, R. D., P. R. Taylor, M. M. Zach and C. W. Heath. "Studies of Human Populations Exposed to Environmental Chemicals: Considerations of Love Canal," in As-sessment of Multichemical Contamination, Proceedings of an International Workshop, Milan, Italy, April 28-30, 1981 (Washington, DC: National Academy Press, 1982) pp. 289-306.

15. "National Death Index. User's Manual," U.S. Department of Health and Human Services, National Center for Health Statistics, Hyattsville, MD (September 1981).

16. "Polychlorinated Biphenyls (PCB) Exposure at Superfund Wastesites," Centers for Disease Control, Center for Environmental Health, Chronic Diseases Division and Clinical Chemistry Division. Unpublished study protocol. (August, 1983).

17. Kleinbaum, D. G., L. L. Kupper, and H. Morgenstern. Epidemiologic Research: Principles and Quantitative Methods. (Belmont, CA: Lifetime Learning Publications, 1982).

18. McMahon, B. and T. F. Pugh. Epidemiology: Principles and Methods. (Boston: Little Brown, 1970).

19. Friedman, G. D. Primer of Epidemiology, 2nd Ed. (New York: McGraw-Hill, 1980).

20. Talbott, E. and E. Radford. "Mortality Experience of a Community Exposed to Low Levels of Radiation." Proposal to be submitted to EPA (1984).

21. Talbott, E., E. Radford, R. Schmeltz, P. Murphy, et al. "Distribution of Thyroid Abnormalities in a Community Exposed to Low Levels of Gamma Radiation," submitted to Am. J. Epid. (1983).

22. Marsh, G. M., J. Costantino, and E. Lyons. "Health Effects of Exposure to the Drake Chemical Company Superfund Site: Morbidity Patterns Among Former Employees," submitted for publication.

23. Logue, N. J., J. M. Fox, et al. "Residential Health Study of Families Living Near the Drake Chemical Superfund Site in Lock Haven, Pennsylvania," Arch. Environ. Health, in press.

24. Lyon, J. L., M. R. Kauber, W. Graft, et al. "Cancer Clustering Around Point Sources of Pollution: Assessment by a Case-Control Methodology," Environ. Res. 24:29-34 (1981).

25. Morgenstern, H. "Use of Ecologic Analysis in Epidemiologic Research," Am. J. Pub. Hlth. 72:1336-1344 (1983).

26. National Technical Information Service. General Catalog of Information Services No. 7. U. S. Dept. of Commerce (1981).

27. Marsh, G. M., D. Schaid, S. Sefcik, B. Miller, et al. "Mortality and Population Data System," unpublished user manual. University of Pittsburgh, Department of Biostatistics (1984).

28. Gittelsohn, A. M. "On the Distribution of Underlying Causes of Death," Am. J. Pub. Hlth. 72:133-140 (1982).

29. Greenberg, E. R., T. Colton and C. Bagne. "Measurement of Cancer Incidence in the United States: Sources and Uses of Data," J. Natl. Cancer Inst. 68:743-750 (1982).

30. Data Bases in SEEDIS, Lawrence Berkeley Laboratory, Computer Science and Mathematics Department. (Berkeley, CA: August, 1982).

31. Merrill, D. W. "Overview of Integrated Data Systems," in Report LBL 15074, Proceedings of the 1982 Integrated Data Users Workshop. (Reston, VA: October 1982, The American Demographics, 1983).

32. "UPGRADE User's Support. A Summary of Data Bases and Data Collections Available Through the UPGRADE System," (Washington, DC: Sigma Data Computing Corp., April, 1980).

33. "UPGRADE User News 3 (Nos. 2-4) 1981 and 4 (No. 1) 1982," Council on Environmental Quality, UPGRADE Project, Washington, DC.

34. McCrea-Curnen, M. G., and E. R. Schoenfeld. "Standard Information on Environmental Exposure Linked to Data on Cancer Patients, with a Brief Review of the Literature," Prev. Med. 12:242-261 (1983).

35. Marsh, G. M. "Proportional Mortality Patterns Among Chemical Plant Workers Exposed to Formaldehyde," Brit. J. Ind. Med. 39:313-322 (1982).

36. John, L. R., G. M. Marsh and P. E. Enterline. "Evaluating Occupational Hazards Using Information Known Only to Employers. A Comparative Study," Brit. J. Ind. Med. 40:346-352 (1983).

37. Neutra, R. "Roles for Epidemiology: The Impact of Environmental Chemicals," Environ. Hlth. Persp. 48:99-104 (1983).

38. Selikoff, I. J. "Commentary, Clinical and Epidemiological Evaluation of Health Effects in Potentially Affected Populations," Environ. Hlth. Persp. 48:105-106 (1983).

39. Cormach, R. M. "Numerical Taxonomy," J. Royal. Stat. Soc. A. 134:321-325 (1971).

40. Kleinbaum, D. G., and L. L. Kupper. Applied Regression Analysis and Other Multivariate Methods (North Scituate, MA: Duxbury Press, 1978).

41. Jones, J. H., L. Jones, J. E. Monson, M. Chapman, et. al. "Numerical Taxonomy and Discrimination Analysis Applied to Non-Specific Colitis," Quart. J. Med. 42:715-732 (1973).

42. Maugh II, T. H. "Biological Markers for Chemical Exposure," Science 215:643-647 (1982).

43. Bloom, A. D., Ed. Guidelines for Studies of Human Populations Exposed to Mutagenic and Reproductive Hazards (White Plains, NY: March of Dimes Birth Defects Foundation, 1981).

44. Lowrance, W. W., Ed. "Assessment of Health Effects at Chemical Disposal Sites," in Proceedings of a Symposium at Rockefeller University, June 1-2, 1981. (New York: Rockefeller University, 1981).

45. "Health Effects of Toxic Wastes," National Institute of Environmental Health Science. Environ. Hlth. Persp. Vol. 48, (February, 1983).

46. Evaluation of Health Effects from Waste Disposal Sites. Fourth Annual Symposium on Environmental Epidemiology, University of Pittsburgh. (Pittsburgh, PA: in press).

CHAPTER 7

FEASIBILITY STUDY TO RELATE ARSENIC IN DRINKING WATER
TO SKIN CANCER IN THE UNITED STATES

Julian B. Andelman and Margot Barnett

INTRODUCTION

In the U.S. Environmental Protection Agency (EPA) 1980 document Ambient Water Quality Criteria for Arsenic [1], a risk
estimate was developed for non-melanoma skin cancer due to arsenic exposure from drinking water based on an epidemiological
study by Tseng and co-workers in Taiwan [2]. This analysis estimated a lifetime risk of 10^{-5} for a lifetime exposure to
drinking water containing 0.025 µg of arsenic per liter. The
present study is an analysis of the feasibility of undertaking
an epidemiological investigation to confirm whether this
Taiwan-based risk estimate is applicable to United States populations.
Several investigations have attempted to determine if there
is a relationship between arsenic in drinking water and skin
cancer in the United States. These will be discussed. The EPA
risk model will be evaluated briefly, as will the baseline prevalence and incidence of such cancer as related to ultraviolet
sunlight exposure. Such baseline cancer rates will be used to
modify the EPA model and apply it in the feasibility analysis.

STUDIES OF U.S. POPULATIONS EXPOSED TO ARSENIC IN DRINKING WATER

The Taiwan study of Tseng et al. [2] showed a dose-response
relationship which is a reasonable basis for a risk estimate of
non-melanoma skin cancer related to arsenic exposure from
water. One of the attempts to determine such a relationship in

the United States was the retrospective analysis of nonmelanoma
skin cancer incidence over a 14-year period in Lane County,
Oregon [3]. No relationship was found between water arsenic
concentrations and skin cancer incidence with about 3700 cases
being observed. In Lassen County, California, although measurements of hair arsenic levels were found to be significantly
elevated in persons consuming well water with arsenic concentrations of 0.05 mg/l, no illnesses associated with arsenic
were found by the health questionnaire survey [4]. In Fairbanks, Alaska, no skin cancer was found in people exposed to
arsenic in water from individual wells [5]. Hair and urinary
arsenic levels showed a dose-response relationship with water
arsenic levels. Urinary arsenic seemed to be a more consistent
indicator of exposure. No differences in the signs, symptoms,
or clinical findings were evident across exposure categories.
Dose levels were low, duration of exposure was short (10
years), and the study population was small (119 exposed). All
of these factors may be related to the negative findings of the
study.

In West Millard County, Utah, a cross-sectional study which
included a physical examination of 249 participants did not
find any association of skin or other disease with arsenic
dose, although a dose-response relationship of hair and urinary
arsenic levels with water arsenic levels was seen [6]. Signs
of arsenic toxicity were found in 6.15% of the exposed population and 2.86% of controls. Cancer mortality patterns did not
differ significantly in the exposed versus control community.

These few epidemiological studies done in the United States
have not involved as high water-arsenic exposures or as large a
study population as that in Taiwan, and their findings do not
demonstrate the influence of such exposure on skin cancer that
was found there. However, because of these population sizes
and lower dose limitations, the lack of such a positive effect
is not necessarily inconsistent with the EPA Taiwan-based risk
model.

NATURE OF THE MODEL

The principal focus of this study is the possible utiliza-
tion of the mathematical predictive model developed by the EPA
to estimate the risk of non-melanoma skin cancer in United
States populations from exposure to arsenic in drinking water.
This model was described in the EPA document Ambient Water
Quality Criteria for Arsenic [1]. The appendix of that document discusses the nature of the model and uses it to develop a
lifetime cancer risk from arsenic ingestion based on the epidemiological study of a rural population in Taiwan [2].

The EPA model utilizes an equation developed by Doll [7]
relating the incidence rate, I, of a site-specific cancer to
the exposure of a carcinogen (herein expressed as concentration

of arsenic in drinking water, C, in mg/l) and the age of the exposed population, t, as follows:

$$I(C,t) = v\beta C^m t^{v-1} \tag{1}$$

where m, v, and β are unknown and adjustable parameters. As discussed in the EPA document [1], this relationship was integrated over time to obtain the cumulative probability density or prevalence, F, which is a function of the same variables as the incident form and is expressed as follows:

$$F(C,t) = 1 - \exp(-\beta C^m t^v) \tag{2}$$

This is a Weibull distribution.

Next, the skin cancer prevalence rates from the Taiwan study were fit to the model using three age groups for males only, namely 20-39, 40-59, and ≥60 or older, used in the model as ages 30, 50, and 70, respectively. Similarly, three concentrations ranges were used: 0-0.29, 0.30-0.59, and ≥0.6mg As/l, as 0.15, 0.45, and 1.2 mg As/l, respectively. These data were fit to a logarithmic form of Equation 2 using least square techniques, and it was judged that there was "an excellent fit having a multiple correlation coefficient of 0.986" [1]. Using the parameters fit to the logarithmic form, Equation 2 was expressed as:

$$F(C,t) = 1 - \exp(2.429 \times 10^{-8} C^{1.192} t^{3.881}) \tag{3}$$

with the parameters in the equation being: $\beta = 2.429 \times 10^{-8}$; m=1.192; and v=3.881.

It was noted that if m were equal to unity, rather than 1.192, then Equation 3 would be "one-hit" in form. Using the Student t test, it was judged that because of the size of the standard error of m, a value of 1.0 could be used for it and this was done. On this basis, the logarithmic form of Equation 2 was again fit to the data and the parameters were redetermined as follows: $\beta = 2.41423 \times 10^{-8}$ and v=3.853, with m=1 being assumed as noted above, the equation being:

$$F(C,t) = 1 - \exp(2.414 \times 10^{-8} C t^{3.853}) \tag{4}$$

A correlation coefficient of 0.971 was calculated. The goodness-of-fit was shown graphically for the logarithmic form of Equation 4 [1]. It showed that for ages 50 and 70, Equation 4 underestimates the prevalence for the lowest of the three concentration ranges; and Equation 4 overestimates prevalence for age 30.

Although not of direct concern in this study, the EPA document developed an estimate of a lifetime probability of skin cancer in the presence of competing mortality using the age-specific incidence rate, Equation 4, assuming a median lifetime of the U.S. population of 68 years of age, v=3.853, β=2.41423x 10^{-8}, and (apparently) m=1. On this basis, the lifetime probability of skin cancer, Q, was related to the arsenic concentration, C, in the well water (as mg As/l):

$$Q = 2.414C/(2.414C + 6.028) \tag{5}$$

At small arsenic concentrations, certainly in the vicinity of 0.001 mg/l and less, Q becomes directly proportional to C, with Q=0.4C. Thus, for example at an arsenic water concentration of 0.001 mg/l, Q equals 4×10^{-4} and, as noted in the document [1], for 0.025 g As/l (2.5×10^{-5} mg/l), the lifetime risk is 10^{-5}.

There are some uncertainties in the development of both the lifetime risk relationship and the prevalence and incidence relationships based on the Taiwan data. Although they appear to be reasonable choices, the three point values of the arsenic concentrations, as well as those for the three age groups are arbitrary, and other choices should lead to somewhat different values for the constants in Equation 4. More importantly, using the incidence, prevalence, and lifetime risk relationships in a form which is first order in arsenic concentration, C (the assumption that m=1 rather than 1.192) may overestimate the risk at the lower concentrations. The overestimation becomes increasingly greater well below the arsenic concentration range from which the relationship was derived. This results from the fact that the concentration term is of the form $C^{1.192}$ with C expressed as mg As/l. Since the arsenic concentrations of interest are considerably less than 1 mg/l, $C^{1.192}$ will be less than C. For example, with C=0.001 mg/l, $C^{1.192}=0.26 \times 10^{-3}$, a factor of 0.26 less than C. At the 10^{-5} lifetime risk level, the EPA calculated that this corresponded to C=2.5×10^{-5} mg As/l, the risk being linear with dose in this range. Using $C^{1.192}$ the risk would in fact be lower than the EPA calculated value by a factor of 0.13. In considering the use of m=1 rather than 1.192, the EPA noted that the standard error of the mean of m was sufficiently large to indicate that these two values could be considered statistically indistinguishable. However, on this basis an even larger value of m is equally valid, perhaps as high as 1.4, which would imply an even lower lifetime risk than that calculated by the EPA.

This analysis indicates that there may be a number of factors in the use and interpretation of the Taiwan data for the purpose of estimating cancer risk that imply a degree of uncertainty that warrants caution in the strict application of these derived relationships. Attention should also be drawn to the fact that the EPA risk estimate was developed from the data for the Taiwan male population, the risks for comparable exposure for females being substantially lower.

This EPA risk model treats prevalence as cumulative incidence, assuming that once individuals in Taiwan developed the disease, they continued to have it for the rest of their lives. If this were not the case in the United States, then prevalence rates here from arsenic exposure should be different (smaller) from those in Taiwan, even if the inherent risks are

the same.   This conclusion is based on the assumption that the
incidence of the disease should be relatable to the exposure.

There is a variety of possibilities that could be envi-
sioned in considering the relative duration of non-melanoma
skin cancer in the Taiwan and United States populations.   As
will be discussed subsequently, the duration of disease for
non-melanoma skin cancer primary lesions in the United States
is typically 2 to 3 years, while in the Taiwan population one
can estimate a range of 8 to 15 years for males and females 40
to 70 years old, using the EPA risk model.   If these substantial differences are correct, this would mean that using the
shorter disease duration here, the prevalence predictable in
the United States from the EPA incident relationship, Equation
1, should be substantially less than that from the direct use
of the prevalent relationship, Equation 4, for a given arsenic
exposure.   This should not, however, affect the lifetime probability of contracting the disease in the United States derivable from Equation 4, since such a probability is independent
of whether the person is treated for the disease.   Also, in
considering the feasibility of conducting an epidemiological
study in the United States, the power of the prevalence study
should be considerably less than that for a similarly exposed
population, such as that in Taiwan, with its expected higher
prevalence rate.

It should also be noted, however, that if some of the
disease in Taiwan had been treated, then clearly the incident
risk there was underestimated, as was, therefore, the lifetime
risk for a given arsenic exposure.   These prevalence-incidence
factors are essential for and will be considered in our assessments of previous epidemiological studies of non-melanoma skin
cancer possibly related to arsenic in United States populations, as well as in the feasibility of a study here involving
exposure to arsenic from drinking water.

UV-B SKIN CANCER STUDIES

In order to assess the occurrence of non-melanoma skin cancer due to arsenic exposure, background levels of skin cancer
should be evaluated.   Insolation and UV-B (ultraviolet light in
the biologically active range of wavelength) exposure vary considerably across regions of the United States depending upon
latitude, geography, and other factors.   When selecting a site
for an epidemiologic investigation of non-melanoma skin cancer
due to arsenic, the possible incremental risk of disease due to
arsenic compared to the background incidence should be evaluated.   If the incremental risk due to arsenic exposure is small
relative to the background risk of disease related to UV-B exposure, then considerably larger populations would be required
to achieve the necessary statistical power to detect the arsenic influence than would be the case in the absence of such a
background incidence.

Several studies have been carried out over the last 10 years in an attempt to assess the public health impact of non-melanoma skin cancer in the United States, and to define the relationship between solar UV-B radiation and skin cancer. An initial incidence survey was performed by Scotto et al. in a six-month period of 1971-72 [8]. The survey covered four areas included as part of the Third National Cancer Survey: Dallas-Ft. Worth Standard Metropolitan Statistical Area (SMSA), Texas; Iowa (state); Minneapolis-St. Paul (SMSA), Minnesota; and San Francisco-Oakland (SMSA), California. Data on incident cases of basal cell and squamous cell carcinomas in Caucasians were collected from dermatologists, pathologists, radiotherapists, and other physicians who diagnose and treat skin cancers. The total population in the survey area was approximately 10 million. Bowen's disease, carcinoma in situ and unknown forms of non-melanoma skin cancer were excluded from the study.

Incidence rates determined by this study were 2 to 3 times greater than had been reported for these areas in the past. Male rates were 2 times greater than female rates. Basal cell carcinoma occurred 3 to 6 times more often than squamous cell carcinoma. Rural populations had a lower risk than urban populations at the same latitude, although this may be due to underreporting of rural cases. The head, face, and neck were the most common anatomical sites for the cancers.

Data from the initial study of Scotto et al. [8] were utilized by Fears et al. [9] to develop models of age and ultraviolet radiation effects on skin cancer. Values for the annual UV exposure index were obtained by use of Robertson-Berger meters placed at airports in the 4 survey areas in 1974. The power function and model fitted to the data will be discussed subsequently. Variations in the observed incidence rates versus the calculated rates using the model were ascribed to possible differences in exposure habits and ethnic differences in skin pigmentation.

A larger scale 1-year incidence survey encompassing 8 areas with a broad geographic and latitudinal range was performed in 1977-78 by Scotto et al. [10]. The methodology was the same as that used in the earlier study by Scotto et al. [8]. The 8 areas included were: Seattle (King County only), Washington; Minneapolis-St. Paul (SMSA), Minnesota; Detroit (SMSA), Michigan; Atlanta (SMSA), Georgia; New Orleans (Metropolitan area), Louisiana; Utah (state); and New Mexico (state). The latitude of the study sites ranged from 30.0 to 47.5 degrees north, and annual UV counts ranged from 101 to 197 (x 10^4) UV-B radiation units. Skin cancer incidence rates were reported by age, sex, race, geographic location, cell type, and anatomical site. Basal cell carcinomas represented 80% of incident cases. A 15 to 20% increase in incidence was seen in those areas which had been included in both surveys. Ten percent of all cases had multiple cancers. A comparison of the incidence rates predicted by the model and those observed is shown in Table 1.

The model mentioned earlier was applied to the data. A plot of log age-adjusted incidence versus UV-B count yields a

Table 1. Observed and Predicted Incidence Rates Per 100,000 of Non-Melanoma Skin Cancer in White Males and Females, Age 50.

Area	UV-B[a] Count	White Male Obs.	White Male Pred.	White Female Obs.	White Female Pred.
Seattle	101	362	272	246	191
Minneapolis- St. Paul	106	334	293	260	202
Detroit	110	241	310	174	212
Utah	147	682	486	357	302
San Francisco	151	409	506	286	312
Atlanta	160	813	554	469	335
New Orleans	176	820	642	454	376
New Mexico	197	660	764	378	431

[a]UV-B = ultraviolet light in the biologically active range of wavelength.

straight line with a positive slope. The model was used to predict the effect of a 1% increase in annual UV-B exposure on skin cancer incidence for each geographic location. The overall result of this estimation was that a 2% increase in skin cancer incidence could result from a 1% increase in UV-B.

Since the predictive model demonstrates the impacts of age, sex, and UV-B exposure on non-melanoma skin cancer on white populations in the United States, all of these factors must necessarily be incorporated into the estimates of expected background incidences when considering a likely study population exposed to arsenic in drinking water. As will be discussed subsequently, a prevalence study will be the focus of the water feasibility analysis, and National Health and Nutrition Examination Survey (NHANES) background prevalence rates will be the principal data base that will be used to account for UV-B and other background effects.

PREVALENCE-INCIDENCE FACTORS

The EPA risk model [1] is based on the Taiwan cross-sectional prevalence study and treats prevalence there as cumulative incidence. A direct concern in the use of the model to determine the feasibility of an epidemiological study in the United States is that the assumption of the EPA model that prevalence equals cumulative incidence may not hold for United

States populations. An estimate of duration of skin cancer
lesions in the United States population based upon the ratio of
disease prevalence, from the National Health and Nutrition
Examination Survey [11], to incidence data from the 1977-78
incidence survey [8] found an estimated disease duration of 2
to 3 years as shown in Table 2. The ratio of baseline prevalence to modeled incidence represents an estimate of duration
of disease in the various age groups.

Table 2. Estimation of Disease Duration in the United States
          Population.

Age	Baseline Prevalence[a] Per 1000		Baseline Incidence[b] Per 1000		Basal Cell Prevalence x 1.25 ÷ Baseline Incidence	
	Males	Females	Males	Females	Males	Females
35-44	4.4	3.8	2.3	1.7	1.90	2.21
45-54	13.4	9.5	5.1	3.1	2.62	3.06
55-64	19.4	14.3	9.6	5.2	2.02	2.74
65-74	33.9	18.3	16.4	8.0	2.07	2.28

[a]From the National Health and Nutrition Examination Survey
 (NHANES) multiplied by 1.25.
[b]From survey of selected regions in U.S. (Scotto, 1980).

     A survey of medical records of 136 skin cancer patients in
Pittsburgh, Pennsylvania, showed an average lesion duration of
1.9 years for males and 2.9 years for females with primary
basal cell carcinomas. Recurrent lesions were more persistent,
with an average duration of 7.6 years in males and 7.0 years in
females. The recurrent cases generally represent 5-10% of all
treated skin cancers. In contrast, an analysis of the prev-
alence/incidence relationship in Taiwan, using the EPA prevalence and incidence models, yields an estimated duration of 8.5
to 14.5 years. Based on these analyses, it appears that the
Taiwan-based prevalence model should not be used directly to
predict prevalence of arsenic-induced skin cancer in the United
States; however, a modified form of this model, taking into
account the likely shorter duration of disease in the United
States, would be appropriate.
     It should also be noted that if the EPA hypothesis for the
Taiwan study is not correct, namely that the observed prevalence rates do not reflect all cumulative incidence, then the

incidence-dose relationship based on the model implies a greater risk for a given arsenic exposure. This indicates, therefore, that there would be greater statistical power in studying
the incidence or prevalence of non-melanoma skin cancer relatable to arsenic exposure in the United States than would be
implied by using the modified EPA model discussed above.

FEASIBILITY ANALYSIS

The feasibility of mounting an epidemiological study of
skin cancer due to exposure to arsenic via drinking water to
confirm the Taiwan-based relationship between exposure and
disease centers upon the ability to locate a community with the
following characteristics:

o A sufficiently high, well-characterized exposure to
  arsenic in drinking water.
o Exposure persisting over a time period sufficient to
  allow a large enough total arsenic dose and to take into
  account the latency period for the development of the
  disease.
o No substantial exposure to arsenic from other sources.
o A large enough population to have sufficient statistical
  power to distinguish differences in skin cancer rates between the exposed and control populations and correlate
  the cancer rates with the exposure levels of the population at risk.

The possible study sites for an epidemiological investigation were determined essentially from an analysis of data in
reports of violations of arsenic limitations specified in the
EPA Interim Primary Drinking Water Regulations applicable to
public water supplies. Hanford City, California, was selected
as a possible study site due to its relatively large population
size and history of repeated violations of the arsenic maximum
contaminant level (MCL) of 50 $\mu g/l$. Because of the well documented and extensive analysis of arsenic in the well water system, it was concluded that the first of the characteristics
listed above is met. It is somewhat more difficult to ascertain whether the second desired feature is met by this site.
From available data, it appears that a maximum of 15,000 people
have exposures of at least 10 years. The reported latency
period for arsenic skin cancer is 10 to 18 years [12,13,14].
Based on an assumed average arsenic concentration of 100 $\mu g/l$
in Hanford City water and 2 l/day consumption of contaminated
water, a person exposed for 10 years would have received a
total arsenic dose of 0.73 g. This is above the lowest reported total doses known to have caused skin cancer in medicinal
applications, such as 0.57 g found by Fierz [13], and 0.144 g
by Neubauer [12].

There is some risk of past exposure to arsenic from the use of arsenical pesticides in the Hanford City area, which would have to be investigated further if a study were found to be feasible based on the last factor listed above. The issue of the statistical power of a study to detect the incremental risk of skin cancer due to arsenic above the background levels of disease is central to the feasibility analysis. In order to examine the issue, several questions should be considered:

o How will possible differences in the prevalence or incidence of skin cancer in the United States versus Taiwan
  affect the feasibility of a study?
o What is the impact of UV-B induced skin cancer rates on
  the feasibility of uncovering an arsenic-related risk
  from water exposure?
o What type of an epidemiological investigation is optimal
  for uncovering a relationship similar to that found in
  Taiwan?
o If it is concluded that a study of the United States
  populations with known exposures to waterborne arsenic is
  not feasible because of limitations of statistical power,
  what size populations and/or arsenic exposures would be
  required to mount a successful study?

The population of Hanford City at the time of the 1980 census, along with that of the 2 neighboring communities of Home Garden and Armona, also high in arsenic exposure, was about 25,000. The various wells ranged in arsenic concentration from less than 10 to 253 μg/l, an estimated typical mixed-well weighted concentration being in the vicinity of 100 μg/l. Although available data would permit a more precise estimate of the latter value, for the purposes of the projected estimate of cancer prevalence from arsenic, concentrations of 50-, 100-, and 200 μg/l will be utilized.

A critical question is what kind of an epidemiological study is appropriate for Hanford City, taking into consideration the availability of information about the arsenic exposure and retrospective medical information relating skin cancer in-
cidence or prevalence in the community. Based on discussions with staff of the Kings County Health Department in the Hanford City area, it was judged that there would not be sufficient information to do a retrospective study of skin cancer in the community. A key point is that there would not be reliable and usable medical records that would define sufficiently past (non-active) cases of non-melanoma skin cancer, nor would a detailed medical history uncover such cases in the survey popula-
tion with the required degree of reliability. For example, it is unlikely that histological analysis of biopsied tissue would be widely available. For these reasons, it was decided that only currently confirmable cases could be the subject of an epidemiological study there.

Due to the variable mixing of well waters within Hanford City, it is unlikely that a dose-response study could be initiated with any degree of confidence. Therefore, it was decided that a cross-sectional prevalence study would be most appropriate along with a low-arsenic exposure control population, such as that of nearby Tulare, California, with a 1980 census population of 22,465. For the purpose of the risk calculation, it will be assumed that the control population is the same size, and has age and sex distributions similar to those of Hanford City. A study protocol would include physical examination and verification of lesions by biopsy. Finally, because the more substantial skin cancer risk is expected in the older members of the population, the risk calculations will be made on individuals older than age 35. Based on the 1980 census, the population distribution for males and females above age 35 is shown in Table 3, our total at-risk group being 9309. For this population distribution, baseline skin cancer prevalence was calculated for males and females using NHANES data as discussed earlier and are shown in Table 4.

Table 3. Population Distribution, Ages 35-85, in Hanford City Area Based on 1980 Census.

Age	Male	Female
35-44	1347	1389
45-54	1018	1100
55-64	898	1098
65-74	675	891
75-84	307	586
Total	4245	5064
Total, both sexes	9309	

The question of the direct applicability of the EPA model based on the Taiwan data was discussed previously. It was concluded that the model should be modified to predict prevalence in the United States due to differences in the ratio of prevalence to incidence between the United States and Taiwan. The modification consists of taking the average ratio of NHANES baseline prevalence to the baseline incidence (Table 2), and using these average values of 2.2 for males and 2.6 for females as multipliers for the EPA excess incidence model based on the Taiwan data. The expected number of skin cancer cases due to arsenic can then be calculated by this "modified EPA prevalence model." It is clear that if, as is unlikely, the United States disease duration rate is much longer, and equivalent to that assumed by the EPA for Taiwan, then the EPA incidence and prevalence models are both directly applicable here. However, the

Table 4. Expected Number of Prevalent Cases When Age/Sex
Specific Rates are Applied to Hanford City Study
Population of 9,309, Ages 35-85.

Calculation Model	Arsenic Concentration µg/l	Expected Number of Cases	
		Male	Female
Baseline prevalence (NHANES)	--	75.8	63.2
Excess prevalence (EPA model)	50	32.7	15.1
	100	64.9	29.9
	200	128.0	59.0
Excess prevalence (Modified EPA model)	50	4.3	2.3
	100	8.7	4.6
	200	17.4	9.2

modified EPA prevalence model will necessarily predict a much
lower prevalence in the United States for an arsenic exposure
equivalent to that in Taiwan. The factor relating the two prediction rates is typically in the range of 3 to 7, but highly
variable with age.

Using both the EPA and the modified EPA prevalence models,
excess prevalence was calculated for each of these arsenic concentrations: 50, 100, and 200 µg/l. The various predicted
prevalence rates are shown in Table 4 for the total populations, male and female, listed in Table 3. It can be seen that
the expected number of cases increases linearly with arsenic
exposure; the male prevalence rates and, hence, expected number
of cases are higher for males than for females; and the EPA
model projects substantially more cases than does the modified
EPA model.

The next question addressed is the statistical power of
such a cross-sectional prevalence study to detect a predicted
difference between the at-risk and control populations. As described by Fleiss [15], the sample size, n, required of each of
these two populations to test the null hypothesis that the proportions of disease in the two populations are equal, using a
one-tailed test at a significance level with power $(1-\beta)$ is:

$$n^{1/2} = \frac{C_{\alpha}(2\bar{P}\bar{Q})^{1/2} - C_{1-\beta}(P_1Q_1) + (P_2Q_2)^{1/2}}{P_2 - P_1} \tag{6}$$

where: $Q_1 = 1-P_1$, $Q_2 = 1-P_2$, and P_1 and P_2 are proportions of disease in populations P_1 and P_2. For a given significance level, α, C denotes the value in standard deviations that cuts off the proportion α in the upper tail of the standard normal curve. For example, if $\alpha = 0.025$, $C_{\alpha}=1.96$. The power of the test is $1-\beta$ and is derivable from Equation 6 for a given significance level. Normally, a minimum power of 80% is satisfactory in such a comparative study to detect differences between the exposed and control population. As shown in Table 5, the EPA prevalence model should have sufficient power at any

Table 5. Power and Sample Sizes for Arsenic Prevalence Studies of Populations Aged 35-85, Assuming NHANES Baseline Prevalence.

	1980 Hanford Area Population		
Calculation Model	Arsenic Concentration μg/l	Power %	Population Size For 10% Power[a]
Excess prevalence (EPA model)	50	84.7	--
	100	100.0	--
	200	100.0	--
Excess prevalence (Modified EPA model)	50	10.6	364,000
	100	19.5	93,000
	200	45.7	24,000

[a]Of 35- to 85-year olds with the same distribution as in Hanford City; $\alpha=0.05$, $1-\beta=0.80$.

of the three arsenic concentrations listed. In contrast, however, using the modified EPA model there is not sufficient statistical power at any of the 3 listed arsenic concentrations. The population size (35-85 years of age with the same age distribution as in Hanford City) required for 80% power to detect arsenic-related prevalence differences between the exposed and

control populations was determined using the same type of calculation of statistical power and the NHANES baseline prevalence rate. These are shown in the last column of Table 5 and
indicate a substantially larger (control and exposed) population than the 9309 in this age range in the Hanford City area.
Thus, it is concluded that the more likely prevalence, due to
arsenic, namely that based on the modified EPA model, would not
be detectable in such a study.

ALTERNATIVE STUDIES

There are two basic study designs which are possible alternatives to the prevalence study of Hanford City discussed
above. They are a prospective cohort study and a multi-city
prevalence study. Retrospective studies are not generally
feasible for populations like those of Hanford City because
historic information on skin cancer would not be reliable.
Furthermore, thoroughly defining the population and tracing the
health status of everybody who had ever lived in such an area
would be a near-impossible task. Of the two alternative
designs, the prospective cohort design would be the more feasible. In power calculations done above, numbers of people would
be replaced by numbers of person-years. Thus, following prospectively the population of Hanford City and an appropriate unexposed population for 10 years would theoretically give a
study with reasonable power, assuming for example, an arsenic
concentration of 100 g/l. The focus would be to detect incidence of the disease rather than prevalence. Generally,
incidence is more useful in researching etiology of a disease
than prevalence. One disadvantage is that it would not be the
most useful measure in a comparison with the Taiwan study
results. Incidence of the disease in Taiwan is not known. It
can be estimated using the EPA model, but the actual incidence
is not known. Comparing United States incidence directly with
Taiwan prevalence would obviously be irrelevant.
There are also practical problems with a cohort study, but
they are not unsolvable. Cohort studies are more expensive
than prevalence studies. It is possible to do a prospective
study of a general population, but they are rarely done because
of the cost. To do the study properly, each member of the cohort (both exposed and unexposed members) must be followed
individually for the duration of the study. The location of
people moving out of town and the vital status of those who
died must be known. Records of current health status of living
members must be maintained. Since skin cancer lesions can
appear and disappear in a relatively short period of time, frequent examinations and/or some other method such as histopatho-
logical verification of diagnosis to assure complete reporting
would be required.
An incidence study would have to define whether multiple or
repeat lesions should be counted as separate cases. Careful

records on each individual would have to be maintained to avoid over- or under-counting numbers of lesions.   Another practical problem in a study lasting several years is apathy.   There would be a loss to follow-up because people drop out due to lack of interest, especially in the unexposed population.   Finally, even a well-run prospective study in Hanford City would fail to yield dose-response information because everybody had approximately the same level of exposure.

The other alternative is a multi-city study in which people from many cities would be grouped into one large study for the purposes of analysis.   To consider the feasibility of such an approach, data were taken from EPA reports of public water supply constituent limit violations to group cities by concentration of arsenic in their water supplies.   When more than one concentration value was present for a single water supply, the values were averaged.   Summary data are given in Table 6.   The midpoint arsenic concentration that was used to calculate power is also given.   To do a rough power calculation, the age and racial distributions of Hanford City were used for all cities.

Table 6. Communities Grouped by Water Arsenic Concentration.

	Water Arsenic Concentration (μg/l)						
	50-99	100-149	150-199	200-249	250-299	>300	Total
Population exposed	14,467	52,000	3,413	--	270	1,364	71,516
Population ages 35-85	5,382	19,326	1,269	--	100	507	26,584
Number of communities	30	19	5	0	2	1	57
Average arsenic concentration (μg/l)	75	125	175	--	275	4,100	--

Expected excess numbers of prevalent cases as predicted by the modified EPA model are given in Table 7.   The expected numbers of baseline prevalent cases predicted using the NHANES data are shown in Table 8.   Two exposure ranges are utilized because there is some question as to whether the city in the greater-than-300 category should be included in any study.   If this city is included, using the calculations discussed above it was found that there is 82% power to detect a difference in

Table 7. Expected Excess Prevalent Cases for 57 Study Cities
Using Modified EPA Model.

	Water Arsenic Concentration (µg/l)									
	50-99		100-149		150-199		250-299		>300	
Age	Male	Female	Male	Female	Male	Female	Male	Female	Male	Female
35-44	.44	.18	2.66	1.08	.24	.10	.03	.01	2.28	.94
45-54	.64	.27	3.80	1.62	.35	.15	.04	.02	3.25	1.39
55-64	.94	.46	5.64	2.72	.52	.25	.07	.03	4.86	2.35
65-74	1.10	.57	6.58	3.42	.61	.31	.07	.04	5.70	2.97
75-84	.73	.55	4.38	3.30	.40	.30	.05	.04	3.38	2.84
Total	3.85	2.03	23.07	12.14	2.12	1.12	.26	.14	19.93	10.49

Table 8. Expected Number of Baseline Prevalent Cases (NHANES)
For 57 Study Cities.

	Water Arsenic Concentration (µg/l)			
	50-299 (56 Cities)		50 to >300 (57 Cities)	
Age	Male	Female	Male	Female
35-44	16.5	14.6	16.8	14.9
45-54	38.1	29.3	38.9	29.8
55-64	48.7	43.8	49.7	44.7
65-74	64.0	45.6	65.3	46.4
75-84	44.8	43.8	45.7	44.7
Total	212.2	177.1	216.4	180.5

skin cancer between exposed and unexposed people. If the city is not included, the power drops to 47%. Sufficient power hinges on a single city whose measured arsenic value is so excessive that it is highly unlikely that it is an accurate measurement of people's chronic exposure in that city. Thus, it is similarly unlikely that it accurately reflects people's chronic exposure in that city.

Besides the issue of statistical power, other problems rule out this study design. The management of what amounts to 57 different studies would be unwieldy. Consistent disease definitions would have to be applied in all cities. Because of the many differences in populations, different appropriate control groups might have to be found for all 57 cities. As the numbers of unknown and unmeasured differences between people increases, the hope of getting an unbiased, unconfounded result dwindles.

The lack of accurate exposure measurements would be another problem in a multi-city study. In Hanford City, it was reasonable to assume that residents had approximately the same level of exposure because water supplies were blended together. Other cities would not have such a system. Within a city there may be areas of very high and very low arsenic concentration. Averaging the available samples of water, which were not random samples, would mischaracterize the exposure of the populations, so that any dose-response analysis would be questionable.

CONCLUSION

It is judged that it is unlikely that a United States population exposed to arsenic in drinking water could be found to provide sufficient statistical power for a practicable epidemiological study to confirm the EPA Taiwan-based risk estimate for arsenic-induced skin cancer. However, this should not be interpreted to imply that such exposures in the United States are not associated with carcinogenic risk.

ACKNOWLEDGMENTS

This research has been funded in part by the U.S. Environmental Protection Agency (EPA) under assistance agreement CR 811173-01 with the Center for Environmental Epidemiology, Graduate School of Public Health, University of Pittsburgh. The research was undertaken by the Center for Environmental Epidemiology at the request of the EPA. Substantial input, guidance, and review was provided by the Center Director, Philip E. Enterline. Other major participants in the study were Richard J. Caplan, Jeanette D. Hartley, James Miller, Magnus Piscator, and Lee Ann Sinagoga.

REFERENCES

1. "Ambient Water Quality Criteria for Arsenic," U.S. Environmental Protection Agency, Office of Water Regulations and Standards, Criteria and Standards Division, Washington, DC, EPA 440/5-80-021 (1980).

2. Tseng, W. P., H. M. Chu, S. W. How, J. M. Fong, S. M. Lin, and S. Yeh. "Prevalence of Skin Cancer in an Area of Chronic Arsenicism in Taiwan," J. Natl. Cancer Inst. 40(3):453-463 (1968).

3. Morton, W., G. Starr, D. Pohl, J. Stoner, S. Wagner, and P. Weswig. "Skin Cancer and Water Arsenic in Lane County, Oregon," Cancer 37:2523-2532 (1976).

4. Goldsmith, J. R., M. Deane, J. Thom, and G. Gentry. "Evaluation of Health Implications of Elevated Arsenic in Well Water," Water Research 6(10):1133-1136 (1972).

5. Harrington, J. M., J. P. Middaugh, D. L. Morse, and J. Housworth. "A Survey of a Population Exposed to High Arsenic in Well Water in Fairbanks, Alaska," Amer. J. Epidemiol. 108:377-385 (1978).

6. Southwick, J. W., A. E. Western, M. M. Beck, T. Whitley, R. Isaacs, J. Petajan, and C. D. Hansen. "Community Health Associated with Arsenic in Drinking Water in Millard County, Utah," Report to the Health Effects Research Laboratory, U. S. Environmental Protection Agency, Grant No. R-804 617-01 (1981).

7. Doll, R. "The Age Distribution of Cancer: Implications for Model of Carcinogenesis," J. Roy. Stat. Soc. A134:133 (1971).

8. Scotto, J., A. W. Kopf, and F. Urbach. "Non-Melanoma Skin Cancer Among Caucasians in Four Areas of the United States," Cancer 34:1333-1338 (1974).

9. Fears, T. R., J. Scotto, and M. A. Schneiderman. "Mathematical Models of Age and Ultraviolet Effects on the Incidence of Skin Cancer Among Whites in the United States," Amer. J. Epid. 105(5):420-427 (1977).

10. Scotto, J., T. R. Fears, and J. F. Fraumeni, Jr. "Incidence of Non-Melanoma Skin Cancer in the United States," DHEW Publ. No. (NIH) 82-2433, National Cancer Institute, Bethesda, MD (1981).

11. Johnson, M. "Skin Conditions and Related Need for Medical Care Among Persons 1-74 Years," DHEW Publication No. 79-1660, U.S. Dept. of Health, Education and Welfare (1978).

12. Neubauer, O. "Arsenical Cancer: A Review," Br. J. Cancer 1:92-251 (1947).

13. Fierz, U. "Katmnestische Untersuchungen über die Nebenwirkungen der Therapie mit anorganischem Arsen bei Hautkrankheiten," Dermatologica, 131:41-58 (1965).

14. Roth, F. "The Sequelae of Chronic Arsenic Poisoning in Moselle Vintners," German Med. Monthly 2:172-175 (1957).

15. Fleiss, J. L. Statistical Methods for Rates and Proportions, 2nd ed. (London: John Wiley and Sons, Ltd., 1981).

CHAPTER 8

USE AND MISUSE OF EXISTING DATA BASES IN ENVIRONMENTAL
EPIDEMIOLOGY: THE CASE OF AIR POLLUTION

Peter Gann

INTRODUCTION

Epidemiologic methods have recently come under increased
pressure to provide critical, decision-making information in
the political, regulatory, and legal arenas.   The quest for
higher certainty, faster results, and lower cost tempts many
epidemiologists to consider the use of plentiful and inexpensive data from existing monitoring networks and surveys.   This
paper discusses some of the troubling methodologic questions
raised by the use of pre-existing data bases.   Since yielding
to temptation might not always be wrong (or at least unprofit-
able), the paper also identifies potential strengths of the approach in studying environment/disease associations.   For purposes of illustration, emphasis is maintained on exposure data
bases and on studies of the effects of ambient air pollution.
Many of the points apply to data bases on health effects and to
studies of other types of environmental exposures.
Epidemiologic data is always highly desirable in making environmental policy decisions, as it is based upon observations
of actual, free-living human populations.   Nevertheless, traditional epidemiologic approaches are strained when applied to
detecting and quantifying small relative risks due to environmental exposure for common, multifactorial diseases.   These
types of diseases or health problems, such as cancer, cardiovascular disease, or adverse reproductive outcome, are precise-
ly where public health concern about environmental agents is
currently the greatest.   Epidemiologic studies are therefore
called upon to become more sensitive, that is, more capable of

detecting small but important risks against a noisy background. It may be helpful to view the epidemiologic study in
this context as analogous to a diagnostic test in clinical medicine, one that attempts to detect a problem in an entire population.

Only a few sources of error can be manipulated in order to
reduce total error and thereby increase the sensitivity of epidemiologic studies. A schema that describes these sources of
error is shown in Figure 1. Sample size can of course be increased to minimize the impact of random errors, a fact which
makes large existing data bases appear more attractive. However, the potentially most damaging source of error in environmental studies is systematic or nonrandom error in exposure assessment. This can be due to the use of crude data with low
validity or to improper modeling of the way in which exposure
"behaves." This type of error diminishes study sensitivity or
statistical power in a broad sense, while conventional power
calculations account only for random error and assume the given
exposure data is correct for each individual in the study.
With the goal in mind of maximizing study power in this broad
sense, the following sections explore several ways of understanding the tradeoffs involved in using existing data bases in
environmental studies.

DEFINING THE EPIDEMIOLOGIC RESEARCH QUESTION

What kind of exposure data do epidemiologists need? The
answer depends on the development of a well-defined research
question in each case. This question (or questions) should
usually be defined before exposure or effect data are selected
and should be based on a biologic model of the exposure-effect
relation that is as explicit as possible. For example, the
question "Does exposure to photo-oxidants affect pulmonary

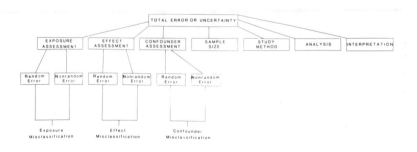

Figure 1.   Sources of error or uncertainty in environmental
epidemiologic studies.

function?" requires refinement towards specifying the photo-
chemical oxidant species of concern, the pharmacokinetics of
ozone in the lung, the population at risk, the use of individual versus grouped data, the temporal aspects of exposure, and
probable concomitant exposures. This specification of the research question is an essential part of determining appropriate
exposure data. As for modeling effect, knowledge concerning
the biology of the pulmonary reaction to photo-oxidants must be
used to hypothesize alterations in specific measurable func-
tions.

Study design also plays a role in determining the requirements for exposure data. An ecological study designed to generate hypotheses does not require the same kind of data as a
prospective study of a cohort. Surveillance studies might tolerate the use of very crude data if the objective is only to
detect major time-space clustering of disease in large populations.

One major aspect of the choice of exposure data involves
the selection of individual versus grouped or aggregate data.
Aggregate data on exposure have often been used in epidemiologic studies of air pollution, since data on individual exposure are usually absent. When health effects are then measured
in individuals, this results in what might be called a "semi-
ecologic" study, in contrast to a full ecologic study, which
contains aggregate data on both exposure and effect. The potential for error introduced by the use of aggregate data on
exposure will be explored further in the next section. With
rare exceptions, existing data bases provide aggregate data
from which individual data can sometimes be derived.

In selecting exposure data, it also helps to specify the
level or type of environment/disease association that is
sought. Four levels or types of association can be examined in
epidemiologic studies, each calling for a different degree of
precision and validity in the exposure data. Every analytic
epidemiologic study generates an exposure-response relationship, even the simplest study which might only compare two
points - exposure and no exposure. Figure 2 illustrates a hy-
pothetical exposure-response curve for an air pollutant; in
this case the "curve" is linear and intersects the exposure
axis at exposure level C. In studies with individual data on
exposure and outcome, each individual will contribute a point,
or individual data will be collapsed into groups to form fewer
points. In studies with aggregate data, obviously each aggregate or group will contribute one point. In Figure 2, level A
refers to a study that seeks "any association" between environment and disease - thus permitting comparison of populations
with maximum contrasts in exposure. This readily allows the
use of more crude data such as might be available in data
bases, since a correct answer can be achieved even if actual
exposures are considerably different from those estimated. On
the other hand, studies that ask questions regarding the shape
(e.g., slope and position) of the exposure-response curve require more finely tuned data on exposure at two or more convenient points (segment B in Figure 2).

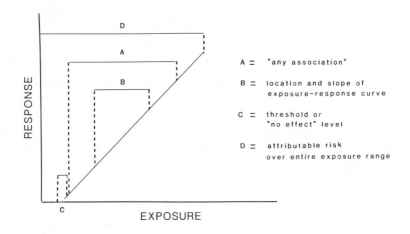

Figure 2.   Four distinct epidemiologic questions and their
            exposure data requirements.

     If the research question concerns determination of a
threshold or "no effect" level of exposure, the investigators
must be able to identify groups or individuals with exposure on
both sides of, and close to, point C. This demands even greater refinement in the exposure data, with less room for nonrandom error.
     Finally, questions concerned with the portion of the total
disease burden attributable to the environmental agent must either identify population exposure across the entire range of
exposure (segment D), or study a representative sample of
cases, as in a case control study.

MISCLASSIFICATION OF EXPOSURE AND ITS CONSEQUENCES

     The epidemiologist must ensure that, to the extent possible
and necessary, individuals are correctly classified with respect to exposure. Failure to correctly classify subjects according to exposure/no exposure or level of exposure (referred
to as misclassification of exposure) will damage the overall
sensitivity of the study [1]. If misclassification is indif-
ferent to health effect status, the contrast between real exposure groups is diluted, study sensitivity is lost, and a
false negative result is more likely. On the other hand, misclassification that assigns higher exposure levels to those
subjects with greater health effects will make a false positive
study result more likely.

Valid information for classifying exposure is not enough. Existing data bases, having usually been collected for some other purpose, rarely contain information on factors that confound or modify exposure, such as cigarette smoking or workplace exposure. Failure to consider these factors, or misclassification of subjects once they are considered, can also contribute to loss of validity in an epidemiologic study [2]. Very slight degrees of misclassification for a strong confounder, such as smoking, can eliminate the power to detect small air pollution health risks. In general, rough but randomly misclassified data on confounders is preferable to no data at all in adjusting crude associations between exposure and effect.

The risk of misclassification of exposure in a data base can be viewed within a framework for describing total personal exposure (see Figure 3). This framework, used by the recent NAS/NRC Committee on the Epidemiology of Air Pollutants, shows the relationship of various levels of exposure measurement to total personal exposure, the best practically available predictor of a health effect in an individual [3]. Data bases whose sole information consists of outdoor pollutant levels at central monitoring stations will give distorted estimates of true total exposure to individuals for many pollutants. The data base will lead to misclassification by failing to account for other sources of exposure or for time-activity patterns that alter true exposure in segments of the study population. For example, use of aerometric data on nitrogen dioxide in epidemiologic studies must be tempered by new knowledge confirming the importance of indoor sources to total personal exposure [4]. For pollutants such as ozone, which have predominantly outdoor sources, community-wide aerometric data might be more valid. The relative amount of time spent outdoors, physical activity, travel between pollutant zones, and the use of air-conditioning might still have to be assessed, depending upon the demands of the particular research question under study. Biologic markers, which are referred to here as measures of exposure obtained in body fluids or tissues, can provide ways to estimate actual dose to target tissues and therefore reduce misclassification [5]. However, few biologic markers for environmental research have yet been developed and validated, and even fewer are likely to be available in large routinely collected data bases, National Health and Nutrition Examination Study (NHANES) notwithstanding.

In trying to formulate the most precise research question possible, we are often stymied by our ignorance of the ultimate chemical species of concern. Hence we often use surrogates, such as SO_2 for sulfur oxides or benzo(a)pyrene for polynuclear aromatic hydrocarbons. Existing data bases, particularly those that are long-lived, are based of necessity on such surrogates. The relationship of the surrogate to the ultimate species of concern, even when it is rather well understood, can vary from individual to individual, due to pharmacokinetic differences, creating another source of error in classifying true exposure.

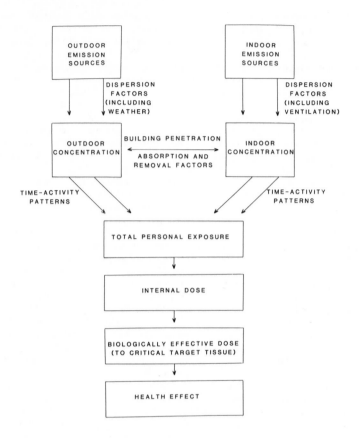

Figure 3.   Framework for exposure assessment.

TEMPORAL AND SPATIAL CONSIDERATIONS

Time relationships are a critical and often-ignored part of the hypothesized biologic model behind the research question. Time lags for acute effects, various dose patterns (e.g., peak versus cumulative), and latency for chronic effects are important features that determine exposure data requirements. In air pollution epidemiology, for example, historical reconstruction of individual cumulative exposure can be important, and yet few studies offer more than categories for subjects such as "lifetime urban" or "previous urban."

As a following example will show, the temporal characteristics of routinely collected data on exposure can easily be manipulated to form reduced or derived data suitable for a particular research question.   Annual average might be appropriate for studies of chronic, cumulative effects, while exposure to peaks or a certain frequency of peaks might be more relevant for other effects.

Each measurement in an air quality data base also represents concentration of a contaminant in a certain spatial volume.   Theoretically, these air parcels range in volume from the air in a person's breathing zone to regional air masses.   Monitoring stations are not sited to coincide with the objectives of an epidemiologic study.   This creates some difficulty in selecting and using data appropriately.   The major national aerometric data system, SAROAD (Storage and Retrieval of Aerometric Data), contains information from sites that are chosen to meet any of four basic monitoring objectives [6]:

o determine the highest concentrations on the network
o determine representative concentrations in areas with the highest population density
o determine the impact of particular sources on ambient pollution levels
o determine general background levels.

Most epidemiologic questions require characterization of exposure to a defined individual or population, which would require a different siting strategy.   Studies of rural populations, which are important with regard to ozone and acid aerosol exposures, are hampered by the emphasis on monitoring urban areas.   Furthermore, combination of data from sites with a variety of objectives can be misleading.

The air parcels actually sampled in SAROAD range from micro scale (several to 100 meters) to regional scale, covering hundreds of kilometers.   The epidemiologist must consider the spatial distribution of pollutant levels during the study design phase, when the model of hypothesized exposure-effect is being developed.   For example, neighborhood scale measurements are often useful because, for certain pollutants, they tend to reflect homogenous exposure to populations large enough to be practically studied, and neighborhood cohorts with contrasting exposure can readily be compared.   The use of regional scale data, on the other hand, is complicated by the difficulty of finding populations with contrasting exposure that are comparable in other respects.

Individuals can move through many of the smaller air parcels in a typical day.   Exposure to some parcels, such as the air at a midtown intersection contaminated with carbon monoxide (CO), might be brief for most persons.   Analysis of such sharp spatial variations in CO levels suggests that use of the monitoring data should be restricted to studies of persons who remain in the area for occupational reasons (e.g., merchants or traffic police), or to studies of very short term effects on persons passing through the area.

ILLUSTRATIONS

There are examples of studies that have avoided many of the aforementioned problems and managed to exploit the sample-size and cost advantages of using ready-made data. Bates and Sizto were able to detect a small ozone effect on asthma admissions - amounting to only about 20 excess admissions per day - essentially by observing 6 million study subjects in Ontario [7]. The availability of easily linked data bases on local pollution levels and on morbidity in a defined population made this possible. Time series analyses similar to this one, in which temporal changes in air pollution are related to acute morbidity events in large populations, could provide very productive epidemiologic applications.

Another study will serve to illustrate some of the decisions involved in using exposure data bases. Portney and Mullahy performed a study which linked SAROAD data to a national sample from the Health Interview Survey (HIS) [8]. Among other health effect variables, this study focused on the number of self-reported restricted activity days due to respiratory disease in the two weeks prior to interview. A total of 3347 subjects were involved in the final analyses. Each subject was matched to 10 air pollution monitors "nearest home" for each of 8 pollutants. Nearest home monitors were determined by matching coordinates of each monitor to the coordinates of the subject's residential census tract. Subjects living more than 20 miles from the nearest monitor were excluded; average distance from a monitor was 4 miles. Rural subjects were further excluded, since census tract coordinates were only available for standard metropolitan areas. Table 1 shows how exposure variables were derived from the monitoring data on ozone and sulfates. It also gives the coefficients for each exposure variable in a multivariate model of acute respiratory disease. Note the three following points:

1. The centroid of the census tract had to be used since coordinates for each individual residence were not available. A subject living at the edge of a tract might well be better characterized by exposure data from the adjoining tract.
2. Ten and 20 mile radii for averaging ozone and sulfate concentrations are arbitrary. Changing the averaging area from 10 to 20 miles changes the model coefficient.
3. Coefficients for sulfates vary more than those for ozone, because sulfate was measured only once every 6 days on average, while ozone was measured hourly; and sulfates have greater spatial variation. Therefore, sulfate variables are more sensitive to selection of exposure proxy terms.

Variables were also derived from SAROAD data for annual averages of ozone and sulfate. Table 2 shows these results. Note that averages were taken over calendar years, rather than

Table 1.   SAROAD Data on Ozone and Sulfate in an Epidemiologic
Study of Respiratory Disease.[a]

Variable Name	Description	Sample Mean	Model Coefficient (t-value)
OZNEAR	Average daily maximum one-hour ozone reading during two week recall period at monitor nearest the centroid of respondent's census tract of residence	0.042 ppm	6.883 (1.97)[b]
OZAV10	Same as OZNEAR but averaged during two weeks over all monitors within a 10-mile radius of respondent's census tract centroid	0.043	6.614 (1.91)
OZAV20	Same as OZAV10 but averaged over all monitors within 20-mile radius	0.044	9.324 (2.41)
S4NEAR	Average 24-hour sulfate concentration during two weeks at nearest monitor	10.876 pg/m^3	-0.005 (0.22)
S4AV10	Same as S4NEAR but averaged over all monitors within 10 miles	10.890	-0.0210
S4AV20	Same as S4AV10 but averaged over all monitors within 20-mile radius	10.700	-0.046 (1.4)

[a]Adapted from Portney and Mullahy, 1984.

[b]t-values of 1.96 or greater approximate $p<.05$ (two-tailed).

over the year actually preceding the health interview.   A person could be interviewed for health effects in January and have exposure measured over the 12 months following.   Considerable additional programming and data management effort would be required to change this situation, but would appear to be warranted based upon the likelihood of month-to-month changes in exposure level and the small size of the effects observed. Note also that coefficients for these annual average variables are higher than those found for two-week averages.   This suggests that the biologic model of exposure-effect that relates

Table 2.   SAROAD Data on Ozone and Sulfate in an Epidemiologic
Study of Respiratory Disease.[a]

Variable Name	Description	Sample Mean	Model Coefficient (t-value)
OZANNR	Average daily maximum one-hour ozone concentration over entire calendar year 1979 as measured at the nearest monitor	0.042 ppm	17.603 (3.18)[b]
OZAN10	Same as OZANNR but averaged over all monitors within 10-mile radius	0.043	19.449
OZAN20	Same as OZAN10 but averaged over all monitors within 20-mile radius	0.044	17.473 (2.12)
S4ANNR	Average 24-hour sulfate concentration over entire calendar year 1979 as measured at the nearest monitor	10.752 pg/m^3	-0.0175 (0.41)
S4AN10	Same as S4ANNR but averaged over all monitors within 10 miles	10.709	-0.0558 (1.34)
S4AN20	Same as S4AN10 but averaged overall monitors within 20 miles	10.588	-0.0765 (1.87)

[a]Adapted from Portney and Mullahy, 1984.

[b]t-values of 1.96 or greater approximate p<.05 (two-tailed).

even acute events such as respiratory infections to long-term previous exposure might be more relevant than a short-term exposure model.

USING DATA BASES TO PREPARE STUDIES

Apart from providing actual data elements, data bases can be used effectively to plan studies by improving the selection of populations for study and estimating required sample sizes.

The importance of using available data in this way can be illustrated by a demonstration of the impact of statistical sampling error during population selection on study power, or the likelihood of missing a significant environmental effect (type II error). In the past, many epidemiologic studies of environmental factors have involved comparisons of two or three geographic areas. Figure 4 shows a typical two-town comparison study that is asking a "C-type" question (see Figure 2): Are there health effects associated with exposure at level A that are not associated with exposure at level B?

URN 1 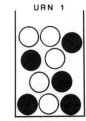   URN 2

Towns with Pollution Level A   Towns with Pollution Level B

Black (B) Balls represent towns with a health effect,

White (W) balls represent those with no health effect,

The probabilities of selecting black or white balls from each urn are:

$P(B_1)=0.5$ $P(B_2)=0.2$

$P(W_1)=0.5$ $P(W_2)=0.8$

Assume Urn 1 has significantly more towns with an effect than Urn 2. Therefore, the null hypothesis (that there is no additional health risk associated with pollution level A) is _false_.

Figure 4.   Sampling error in the selection of towns for a health study.

Each ball represents a town with either exposure level A (urn 1) or level B (urn 2). The investigators will blindly select one ball from each urn for their study. Level A, unbeknownst to the investigators, does have significantly more towns with health effects than level B. Nevertheless, the chances of missing this phenomenon (a false negative result) based on town selection alone are 60%, since only selection of a black ball from urn 1 and a white one from urn 2 will yield a positive result. These probabilities are shown in Table 3.

Table 3.   Risk of Type II Error (Incorrectly Accepting the Null Hypothesis) Based on Town Selection Alone.

URN 1

		B	W
URN 2	B	Outcome a: Accept null	Outcome b: Accept null
	W	Outcome c: REJECT NULL	Outcome d: Accept null

$P(a) = P(B_1) \cdot P(B_2) = 0.1$ Type II error

$P(b) = P(W_1) \cdot P(B_2) = 0.1$ Type II error

$P(c) = P(B_1) \cdot P(W_2) = 0.4$ Correct

$P(d) = P(W_1) \cdot P(W_2) = 0.4$ Type II error

Overall chances of type II error = 0.1 + 0.1 + 0.4 = 0.6 (60%).

More thorough review and careful use of available monitoring data could be used to better characterize small exposure differences between balls in each urn, creating more urns from which to make selections. Additionally, the use of widespread monitoring data can stock each urn with more balls, making it more practical to select more towns. In many cases it is preferable, if a budget permits study of 2000 subjects, to select 20 towns with 100 subjects each rather than 2 towns, each with 1000 subjects. This concept was used recently in a major French study, which compared subjects from 28 towns or urban districts [9].

CONCLUSIONS

This paper presents some aspects of data bases on pollutant concentrations that must be critically examined before such data bases are applied in epidemiologic studies.

Existing data bases have a definite place in environmental epidemiology and we have only begun to explore their utility. Data bases can greatly reduce the cost of studies and can provide large sample sizes with enormous statistical power. However, attempting to achieve greater study sensitivity by increasing sample size alone can be self-defeating if data quality suffers. Data bases do not always have to be used as a source of variables for analytic studies - they can also be used to tell us what problems are important, whom to study, and how many to study. They are particularly useful in surveillance or outbreak detection systems, where relatively crude data may suffice. Linkage of personal data between exposure and effect data bases must be improved, in order to fully exploit cost and surveillance advantages.

The use of existing exposure data for finely detailed research questions - such as those involving very low-level exposure and chronic disease - will be sharply restricted due to the need for more highly customized exposure variables and a lower tolerance for misclassification. Parallel data sets on multiple exposures or confounding variables might also be necessary. Efforts at modeling exposure (as in individual exposure models) can extend the applications of routine data.

Finally, since many of the most important questions in environmental health concern chronic disease, data bases must be improved to provide better coverage of long-term exposure. Routine air monitoring, for instance, has only been available across most of the United States since about 1970. In the future, historical reconstruction of exposure based on such monitoring over a lifetime will become more feasible. Further improvements in the design of periodic health surveys, such as follow-up of the same people rather than complete resampling, will provide additional opportunities. Modifications in current exposure data bases to provide data more consistent with epidemiologic research questions should certainly be considered, in spite of potential expenses.

ACKNOWLEDGMENTS

The author wishes to acknowledge John Bailar, David Bates, Maureen Henderson, and Paul Portney for their stimulating discussion on these topics. This work was funded in part by the United States Environmental Protection Agency under Contract 68-02-4073 to the National Academy of Sciences/National Research Council.

REFERENCES

1. Copeland, K. T., H. Checkoway, A. J. McMichael, and R. H. Holbrook. "Bias Due to Misclassification in the Estimation of Relative Risk," Am. J. Epidemiol. 105:488-495, 1977.

2. Greenland, S. "The Effect of Misclassification in the Presence of Covariates," Am. J. Epidemiol. 112:564-569, 1980.

3. Epidemiology and Air Pollution (Washington, D.C.: National Research Council, Committee on the Epidemiology of Air Pollutants, National Academy Press, 1985.)

4. Spengler, J. D., and M. Soczek. "Evidence for Improved Ambient Air Quality and the Need for Personal Exposure Research," Environ. Sci. Technol. 18:268A-280A, 1984.

5. Gann, P. H., D. L. Davis, and F. Perera. "Biologic Markers in Environmental Epidemiology: Constraints and Opportunities," in Proceedings of the SGOMSEC 5 Workshop in Mexico City, August, 1985. In press.

6. "Network Design for State and Local Air Monitoring Stations (SLAMS) and National Air Monitoring Stations (NAMS)," Federal Register, Title 40, Protection of Environment, Appendix D, p. 122-145, pt. 58, (1979).

7. Bates, D. V. and R. Sizto. "Relationship Between Air Pollutant Levels and Hospital Admissions in Southern Ontario," Can. J. Pub. Health. 74:117-122, 1983.

8. Portney, P. R. and J. Mullahy. "Urban Air Quality and Acute Respiratory Illness," J. Urban Econ. In press.

9. Groupe Cooperatif PAARC. "Air Pollution and Chronic or Repeated Respiratory Diseases: II. Results and Discussion," Bull. Eur. Physio-pathol. Respir. 18:101-116, 1982.

CHAPTER 9

OPENING AND CONTROLLING ACCESS TO MEDICARE DATA

Glenn J. Martin

Medicare files contain information on more than 95% of the aged individuals in the United States. This information includes the individual's Social Security account number; name and address; state, county, and zip code of the individual's residence; age; sex; and race. Detailed information on the health services paid for by Medicare is in the files. This includes information on diagnosis, hospital admissions and discharges, and provider identity. Medicare files have been described as a tremendous potential resource for health research, including epidemiologic studies. Opening access to these files was, therefore, regarded as being of major importance to health researchers.

Data on identified individuals in the Medicare files, however, are protected by the Privacy Act and Section 1106(a) of the Social Security Act. Access to individually identifiable data has always been permitted for employees, contractors, state agencies, and others for program purposes, without the necessity of obtaining the individual's consent. Release to outside research organizations was allowed but only for research funded by the program and directly related to program purposes. The laws permitted other releases. For example, releases for purposes that are compatible with the purpose for which the data were collected were allowed without requiring the individual's consent under the "routine use" provision of the Privacy Act. Nevertheless, as a matter of policy, individually identifiable data were not released during the Medicare program's first 10 years except for purposes directly related to the program.

Change in this policy was initiated as a result of a request in 1977 by Dr. Thomas Mason, National Cancer Institute

(NCI) epidemiologist. Unaware of the longstanding policy pro-
hibiting access for non-Medicare program purposes, Dr. Mason
requested that the Health Care Financing Administration (HCFA)
furnish the names and addresses of Medicare beneficiaries who
would be contacted by NCI (or its contractor) and asked to volunteer to be interviewed. They were to serve as part of a
group to be compared to bladder cancer cases with respect to
the use of artificial sweeteners.

Under the policy that had been followed while Medicare was
administered by the Social Security Administration (SSA), the
request would have been denied. However, the Medicare and
Medicaid programs had just recently been separated from SSA at
the time of Mason's request, and the policies SSA had established were routinely being re-examined to determine if they
were appropriate for the new organization.

Consideration within HCFA and SSA was lengthy and intense.
On the one hand, the confidentiality of data collected under
Social Security programs had always been protected by limiting
the risk through severe restrictions on access. Permitting
access only for purposes essential to the operation of the programs minimized the risk. That is, the only risk that would be
incurred would be that which could not be avoided and still
carry on important program purposes.

On the other hand, it was argued that the basic purpose of
the Medicare program was not just to provide protection to
beneficiaries against the costs of health services. Payment
for services had no special merit by itself. Payment was made
for services so that beneficiaries would have access to services. Those services, in turn, are of value because they may
intervene in the natural course of a disease or injury.

NCI research, of course, is directed toward seeking causes
of disease or factors associated with disease, with the possibility that this would lead to means of intervening in the natural course of the disease. Therefore, it was maintained that
intervention in the natural course of disease was the ultimate
objective of both NCI research and the Medicare program. Sharing a common objective provided the basis for the compatibility
of purpose required in the Privacy Act for the "routine use"
provision.

Actually, data did not have to be released to NCI under
this provision. The Privacy Act also contains a provision for
release within an agency to employees "who have a need for the
record in the performance of their duties." Since all Health
and Human Services (HHS) employees are considered to be in the
same agency for purposes of the Privacy Act, data could be released by HCFA employees to NCI employees when NCI employees
"need the record for performance of their duties." The more
restrictive requirements of the routine release outside the
Agency were used in considering NCI's request, because if release outside the Agency could be justified, it would be difficult to refuse to release the data within the Agency.

It was also pointed out that the Medicare population was
likely to benefit by NCI research, since the cancer being studied had greater incidence among the aged than among younger

populations. Finally, the Secretary of HHS through NCI was obligated by the law authorizing the bladder cancer study to contact aged persons to seek their participation in the study. Intrusion into the privacy of the aged had to happen. With use of Medicare files, the intrusion could be reduced, since it could be limited to persons who were the right age and sex. Otherwise, random digit dialing would be relied upon with a considerably larger number of contacts being required to obtain the participation of the persons necessary to fill each age, sex, and geographical classification.

In April 1978, the HCFA Administrator agreed to release the data to NCI, and the SSA Commissioner concurred. Access was provided. NCI requested assistance for three additional studies over the next two years. HCFA agreed to assist these studies with names and addresses of beneficiaries for comparison group purposes and SSA concurred. Approval of these requests provided support for additional requests of the same kind from NCI, but the policy reflected in the approvals still could be changed without recourse to any formal procedures.

A more permanent basis to this opening of access was provided by SSA with publication of its revised confidentiality regulations in April 1979. The regulation stated: "We will also disclose information under appropriate circumstances for epidemiological and similar research. We consider this health-related activity to be a compatible purpose, since it may help prevent or lessen diseases, and this may reduce the need for benefits under health maintenance programs."

The key is that "epidemiological and similar research" was declared to be a compatible purpose. HCFA followed suit shortly thereafter by amending the systems notices published in the Federal Register to add a new research "routine use" to all of its major data systems. It provided that disclosure could be made "to an individual or organization for a research, evaluation or epidemiological project related to the prevention of disease or disability, or the restoration or maintenance of health...."

Several conditions were placed on such disclosures. Release was limited to projects that were of "sufficient importance to warrant the effect and/or risk on the privacy of the individual" and with respect to which there was a "reasonable probability that the objective for the use would be accomplished." It was also necessary that the project could not be reasonably accomplished without the disclosure. Finally, several requirements were placed on the recipient of the data regarding protection of the data and further disclosure (see Health Insurance Master Record, Federal Register, Part III, Department of Health and Human Services, October 13, 1983, pp. 45719-20).

Note that up until this publication, HCFA had made no releases for non-Medicare purposes except to NCI. Also note that neither the SSA regulation nor the HCFA research routine use limited disclosure by organization. Release was not limited to NCI, to the U.S. Public Health Service, to HHS, nor to other federal agencies or their contractors or grantees. Individuals and organizations outside of the federal government might also have access. The purpose of the project, the importance of the project, and the soundness of the research design were the determining factors.

Since Dr. Mason's request for NCI's study of artificial sweeteners and bladder cancer, HCFA has provided data for more than 30 similar projects. Geographical data on the beneficiary's residence made it possible to select population samples from areas covered by National Institute of Health cancer registries. Beneficiaries were selected from the Minneapolis/St. Paul area to be compared to persons with kidney cancer for an NCI study of kidney cancer related to the consumption of caffeine. Incidence of kidney cancer was high in this area and thought to be related to high consumption of strong coffee by the Scandinavian population in the area.

Samples for comparison groups were furnished in the Texas gulf coast area because of the high incidence of respiratory cancer which researchers suspected to be related to the presence of the petrochemical industry in the area. Samples in the Sun States were provided for a skin cancer study in relation to solar ultraviolet exposure. Phenoxy herbicides exposure was the basis for samples in rural areas in Washington State. Intake of selenium was the focus of a study in the Rapid City, South Dakota area. Residence in Bronx, New York, was the basis for a sample selected for a study of a health maintenance organization in that area.

Other samples were based on the sex of the individuals, as well as geographical location. A sample of males in ten states was provided for a study of breast cancer in males. Female beneficiaries were selected for comparison purposes for a NCI study of lung cancer in women in New Jersey. Samples have been provided to components within HHS other than NCI, including the National Institute on Aging, Centers for Disease Control, National Center for Health Statistics, National Center for Health Services Research, National Institute of Occupational Safety and Health, and the National Heart, Lung, and Blood Institute.

Use of HCFA files in locating individuals has also become important. For some studies, HCFA has furnished the address of the beneficiary, and for others we have provided information on the vital status of individuals. HCFA's first release of beneficiary addresses outside of the Department involved a large location exercise. In fact, the study represented the first release of beneficiary addresses to a private organization for a purpose not directly related to Medicare.

In February 1981, the Johns Hopkins School of Hygiene and Public Health requested HCFA's assistance in locating shipyard workers who had been exposed to low-dose radiation while repairing nuclear submarines. The study was funded by the Department of Energy in cooperation with the U.S. Navy. About 110,000 radiation-exposed workers had been identified, as well as 110,000 similar but non-exposed shipyard workers. Some were currently employed in shipyards, and some had left recently and could be easily located with telephone directories and other public sources. Location of those aged 65 and over was of greatest concern. Johns Hopkins had Social Security numbers for the vast majority of workers; as a result, HCFA had unusually good success, furnishing 79,844 addresses of study subjects.

Release of addresses to Johns Hopkins easily qualified under the new research routine use HCFA had published in the Federal Register. The research was important. It was important not only to the exposed workers, but to other individuals exposed to low-level radiation. It was controlled by two other federal agencies through a contract with Johns Hopkins. Privacy Act requirements applied as effectively as they would had the contract been with HCFA. Also important, the beneficiary had a stake in the study. Its findings could establish whether the individual had need to be concerned with his radiation exposure at work.

HCFA also assisted the National Center for Health Statistics in locating aged participants in their National Health and Nutrition Study. Addresses were furnished for participants who had moved without providing a forwarding address. Addresses were furnished for a follow-up study of Seventh Day Adventists, designed to establish the relationship between life-style and longevity. About 1000 women from a cohort of 8000 women in a cervical cancer study of women treated at eight different hospitals had been lost to follow-up. HCFA provided addresses for those who could be found in Medicare files. A study of women irradiated for benign gynecological disorders was helped by trying to find the addresses for 5354 women. HCFA also searched for about 1100 World War II veterans exposed to hepatitis.

Vital status has been furnished for several studies, including a study of 10,000 members of the American Chemical Society. For deceased individuals, the state of residence is provided so that death certificates can be obtained with cause of death information from state authorities.

It is evident that access to HCFA files has been made available to a wide variety of health research projects. But that access has been carefully controlled to limit it to projects that have been determined to be important, soundly designed, and sufficiently financed. All but two of the foregoing projects helped were federally financed, which means that a federal agency was responsible for monitoring the project and assuring that the provisions of the Privacy Act, a federal law applying to all federal gencies, were followed and properly enforced. Systems security arrangements were also required.

On two occasions, as indicated above, data were furnished for privately funded studies. Data were released to the Rand Corporation for a study of physician practices which was funded by the Robert Wood Johnson Foundation. However, release of the data was made contingent upon Rand signing an agreement granting HCFA's Office of Research and Demonstrations a role in following the course of the study. The other study was funded by the American Cancer Society and NCI reviewed the research protocol and recommended HCFA's assistance.

For each assisted study, HCFA required the submission of the research protocol, review panel approval of the protocol, funding documentation, and human subjects approval where necessary. Each project involving beneficiary contact was subject to review by the HCFA Administrator. When a beneficiary sample was provided, a letter from the Administrator explaining the project, HCFA's assistance, and the right of the beneficiary to refuse to participate at any time without effect on his Medicare benefits preceded any contact by the research organization. As a practical matter, the beneficiary could notify HCFA and stop any contact. Very few complaints of any kind have been received from beneficiaries - less than 20 from all of the studies. On the contrary, researchers have reported extremely high favorable response from beneficiaries.

In assisting research projects, HCFA attempts to limit the intrusion into the beneficiary's privacy to that essential for the project, and examines each request for possible means of reducing the risk to the beneficiary's privacy. Participation in the aforementioned studies, of course, is voluntary, but being asked to volunteer by mail is still an intrusion. HCFA attempts to avoid multiple contacts by a project by using different sample selection criteria when a request is made for participants from a geographical area previously sampled. Possibly, a different terminal digit in the beneficiary's Social Security number might be used in selecting the sample.

Whenever it is feasible, consent of the beneficiary is required. In clinical trials where the population has already been identified and is being seen, consent is always required. Consent forms were furnished by the Department of Labor (DOL) for its study of cohorts of workers who had been exposed to asbestos at the worksite and were being examined periodically. Individual consents were also obtained for a DOL study of workers exposed to cotton dust. DOL wanted Medicare payment data on the workers. Under the Privacy Act, the individual has the right of access to his records and can consent to a copy of his record being made available to another party. Nevertheless, HCFA requires documentation that the individual knows to whom he is consenting to release his records, for what purpose, and the period of time involved.

Frequently, the research does not require identifiable data. Medicare payments, hospital stays, or similar data may be needed, and the actual identity of the individuals may be irrelevant to the study. In such cases, HCFA may delete the beneficiary's name and address and Social Security account number from the file and bind the researcher to a promise not to make any effort to deduce beneficiary identities nor permit anyone else to do so. Under section 1106(a) of the Social Security Act, imprisonment up to one year may be applied to violations of conditions under which data are released by HCFA. (No violations have ever been reported.) Under such releases, beneficiary privacy remains intact. No real person ever comes to know anything about the individual from the file unless the release terms are violated.

Recently, to meet the needs of the American Hospital Association and similar interests, HCFA developed a file based on data from its file of Medicare data on a sample of Medicare beneficiaries discharged from short-term hospitals. After lengthy consideration and discussions, a file called the "Modified MEDPAR File" was developed. It contains detailed information on the services, charges, and diagnoses of beneficiaries in the sample. Protection of beneficiary privacy is provided by deleting all of the data elements likely to permit the identity of a particular beneficiary to be deduced. Provider identity is included in the file, but it was agreed that beneficiary names, addresses, residence location (except to note if same state as provider), sex, race, or age (five-year age intervals were included) would not be needed. To buttress this protection, recipients are required to sign an agreement to protect the data from any effort to deduce beneficiary identity.

As is evident, HCFA has opened access to its data files. However, we believe that the protective procedures we use result in minimal intrusion into the privacy of beneficiaries, an intrusion which almost all beneficiaries, by their willingness to participate in the studies, appear to believe is fully warranted by the benefit to themselves, other beneficiaries, and the public generally now and in the future.

There are other federal data sources that could be helpful to health researchers, but access to such files is denied, even to other federal agencies. For several years, consideration has been given to legislation that would permit sharing of data among a few federal agencies, but continue the prohibition against all outside releases. It appears that the latest form of these proposals, the "Federal Statistical Records Act," has failed to win the support of the principal agencies: the Census Bureau, the Internal Revenue Service, and the National Center for Health Statistics. The primary concern of each of these agencies appears to be that legislation authorizing sharing of data among them might seriously weaken the public support their data collection efforts currently enjoy. That is, individuals might become reluctant to fully participate in providing information to each of these agencies separately if they knew the information might be made available to another agency for another purpose. Their concerns are undoubtedly valid and real. Nevertheless, I believe that it is important to all of us that effort be continued to find ways to share the data without seriously eroding public cooperation.

Thus far, HCFA generally has been able to assist qualified requesters by providing access to its files. This assistance has always been contingent upon the availability of sufficient resources. To date, help has been provided without interference in HCFA's essential activities. But any substantial growth in this assistance, of course, would require careful evaluation of the impact on resource availability.

DISCLAIMER

The work described in this chapter was not funded by EPA and no official endorsement should be inferred.

CHAPTER 10

DRINKING WATER QUALITY DATA BASES

Nancy W. Wentworth, Kaiwen K. Wang, and James J. Westrick

INTRODUCTION

In recent years, there has been increasing interest in the effects of drinking water quality on human health. Researchers are attempting to link exposure to contaminants in drinking water with illness in humans. To do this, researchers must have access to information on concentrations of contaminants in drinking water and rates of illness in the consumers of the drinking water. Also, if the focus of the study is a chronic disease, then historical information on changes in concentrations must also be available. Unfortunately, there will never be enough data available to meet all these needs; the largest data bases are not structured to meet research needs. These data bases were developed to meet regulatory and enforcement requirements. There are, however, recently developed data sets which contain more of the information of interest to epidemiologists.
Information on water quality is maintained by four groups which have different responsibilities and needs:

o The Federal Government - The Environmental Protection Agency (EPA) develops federal drinking water regulations and provides oversight and assistance to state drinking water programs

o State Governments - Responsible state agencies implement and enforce state codes which regulate drinking water in accordance with provisions of the Safe Drinking Water Act (SDWA)

o Public Water Systems - The water systems' managers maintain information needed to manage the systems on a day-to-day basis, to plan for future needs within the water

service areas, and to meet any data requirements placed
on them by the responsible state agency

o The Water Supply Industry - The industry maintains information on its constituents so that it may effectively
represent them in technical and regulatory matters

Each of these groups' data sources, uses, and availability is
different, and will be presented in detail in the following
discussion.

THE ENVIRONMENTAL PROTECTION AGENCY

The Safe Drinking Water Act (SDWA) became law in 1974. The
Act initiated the national drinking water program and gave EPA
the responsibility for establishing enforceable, health-based
concentration limits called Maximum Contaminant Levels (MCL),
and schedules for monitoring and reporting the results of the
monitoring of the contaminants. These regulations are contained in Title 40, Code of Federal Regulations, Part 141, the National Interim Primary Drinking Water Regulations (NIPDWR).
These regulations apply to "public water systems." These are
water systems, regardless of public or private ownership, which
routinely serve 25 or more people or 15 or more service connec-
tions on a daily basis. Within this group, there are "community water systems," which serve year-round residential populations, and "non-community water systems," which serve transient
populations (e.g., gas stations, campgrounds, etc.).
The Act also allows EPA to delegate primary enforcement
responsibility (primacy) to states to give them day-to-day
responsibility for assuring that the statutory and regulatory
requirements are met by all the federally defined regulated
systems. At this time, 54 states and territories have received
primacy; EPA retains authority in the remaining three states
and territories. Each state and territory must provide EPA
with information on the water systems under its jurisdiction
and the quality of the water served by the water systems. Requirements for delegation and reporting are contained in 40 CFR
Part 142, the NIPDWR Implementation Regulations.
The data developed and submitted by the states, and by EPA
where no state is delegated the responsibility, are stored in
EPA's automated data system, the Federal Reporting Data System
(FRDS), which was established in 1978. The system contains information on 59,000 community water systems and nearly 150,000
non-community water systems. There is some variation in the
quantity of the data submitted by each state, but all systems
have unique identification numbers, water source category (surface, ground water, etc.), the population served, and a commu-
nity/non-community indicator. Water quality information in
FRDS is limited to data collected since 1976, contains information only on contaminants included in the NIPDWR, and only includes information for systems which exceed the standards. A

list of regulated contaminants on which exceedence information
is available from FRDS is contained in Table 1. Additional
information on the data available from FRDS is presented later
in this paper.

Table 1.   Contaminants and Indicators of Contamination Regulated
           by the National Interim Primary Drinking Water
           Regulations.

Total coliform	Arsenic	Radium-226, Radium-228
Turbidity	Barium	and gross alpha particle
	Cadmium	radioactivity
Endrin	Chromium	
Lindane	Fluoride	Beta particle and photon
Methoxychlor	Lead	radioactivity from man-
Toxaphene	Mercury	made radionuclides
2,4-D	Nitrate	
2,4,5-TP	Selenium	
Total trihalomethanes	Silver	

    Other EPA data collection efforts have focused on special
needs: quantifying ground-water contamination by synthetic
organic contaminants, developing national estimates of the occurrence of inorganic or radiological contaminants, or attempting to predict the occurrence of contaminants geographically.
These studies are statistically based, and do not contain data
for all water systems. Generally, the sample involves between
1000 and 1500 systems, and is stratified by system size and
primary water source type; geographical stratification is also
used in cases where the survey includes contaminants which are
found only in certain regions of the country. For these studies, the data set contains the system identifier, the analytes
under consideration, and the analytical results. Additional
information on analytical methods used and associated data
quality indicators (e.g., analytical precision and accuracy)
may be available from the study managers.

STATE DRINKING WATER AGENCIES

    State drinking water agencies have historically had author-
ity to control drinking water quality using their general
health and well-being statutes. The various communities, housing developments, food service establishments, etc., were all
inspected at some frequency based on the statutes and rules of

the specific state. Passage of the SDWA consolidated the authorities and regulations, and tended to standardize the regulatory program within each state by putting all of the systems under consistent regulations.

The states have maintained a more direct relationship with the water systems than has EPA, particularly where the state has primacy. The states have been working with individual water suppliers in a technical assistance role for many years. For these reasons, state files on individual water systems contain a significant amount of detailed information on the systems. The files often contain the analytical results of the monitoring conducted for the system by the state before the passage of the SDWA, and the results of the routine monitoring conducted by the system since passage of the Act. The NIPDWR require the suppliers to submit all the results of monitoring to the state, whereas the state must only submit violations of the NIPDWR to the EPA.

State files may also contain other technical information that is useful in epidemiological studies. A water supplier must provide the state with a complete analysis of the water quality of a water source before it can be put into service. In the past, a "complete" analysis would include only the traditional contaminants listed in the NIPDWR and a few other contaminants which impart an objectionable aesthetic character to the water at high concentrations (e.g., iron, manganese, etc.). Most states require suppliers to submit engineering plans and specifications for review before construction can begin on a new system or a substantial expansion or upgrading in an existing system. These engineering documents and source analyses may be available to help define when changes in water quality may have occurred due to use of new water sources or the installation and operation of water treatment facilities.

The engineering and technical files (plans and specifications and reports on site visits and inspections) are usually kept in written form, and may be located in the state's central office or in district or regional offices. Determining which of these offices has the files of interest may prove to be difficult.

The state may keep its water quality compliance information in an automated system, a manual file (ranging from a shoebox to a folder system) or any system in between. The automated systems are structured to contain information on the water system's physical facilities (water sources, types of treatment in place, etc.), the monitoring requirements for the system (analytes and frequency of monitoring), and the analytical results (including the date of sampling or the compliance date). Some systems also maintain information on the analytical method used for the analyses and on the laboratory which conducted the analyses. Also, some states regulate contaminants which are not subject to federal regulation; these data would be available from the state files. Access to any of this information must be arranged through the specific office which has the information.

PUBLIC WATER SYSTEMS

The 200,000+ public water systems bear a regulatory respon-sibility to maintain certain records on the quality of the water served to the water users; they must maintain records of the results of the monitoring required by the NIPDWR. The water suppliers, particularly those whose systems supply a large number of users (more than 5000 people), also maintain a significant amount of information on the physical system (lengths and types of pipe in the water distribution system, plans and specifications for all facilities, etc.), the water treatment processes (raw water quality, amounts and types of chemicals used, finished water quality, operating information on the treatment processes in use, etc.), information on water quality in the distribution system (microbiological contami-nants, corrosion and disinfection by-products, etc.), and fi-nancial and operating characteristics (rate schedules, depre-ciation schedules for facilities, records of water use, etc.).

As noted, the suppliers maintain a wide range of informa-tion which is necessary to manage the system efficiently. Most of the information is maintained in "hard copy" files, although larger systems are automating (or recording on cross-referenced microfiche) much of the physical system data and the operating, financial, and water usage records. Access to the information must be arranged through the manager/owner of the water sys-tem. It should be noted that smaller water systems, particu-larly the non-community systems, are less likely to maintain any records on the system; they do not consider water supply to be their primary business, and therefore do not maintain any business records on it.

THE WATER SUPPLY INDUSTRY

The American Water Works Association (AWWA) is a major organization which represents water suppliers, state and fed-eral regulators, researchers, water users, equipment manufac-turers and sales representatives, and anyone who is interested in drinking water. The Association has recently developed a data base containing information submitted by the largest water suppliers in the country. The data base was created to allow AWWA to better represent its members in regulatory affairs and as a method of identifying technical or research needs within the water supply industry. The data base contains information on water sources, raw and treated water quality, physical fa-cilities (including source collection, treatment, storage, and distribution), and rate and financial information for each sys-tem. At this time, the data base contains information from over 400 water systems.

STRUCTURE OF THE FEDERAL REPORTING DATA SYSTEM

Inventory

The Federal Reporting Data System (FRDS) is EPA's automated system for managing information on the public water systems regulated under the SDWA. The system contains inventory information on over 200,000 public water systems. The elements for which there are data for each system are:

o Public Water System ID - A unique identifier for each system
o Population Served - Average daily population served by the system
o Source - Information on the various sources of water available to the system (grouped by surface, ground, and purchased sources)
o System Type - Whether the system is a community or non-community system

Additional elements for which there may be data are:

o Owner Name/Address - Name and address of the system's owner (which may not be located near the actual system, particularly in privately owned systems)
o Plant Name/Address - Name and address of the system
o Location - Latitude and longitude of the sources, water entry points to the distribution system, etc.
o Treatment - Treatment units in place at each source or facility

Information on each of these data elements may be present for each system; the inventory files are more complete for the 59,000 community systems than for the 150,000 non-community systems.

Violation Files

Information on violations of the maximum contaminant levels specified in the NIPDWR is stored in FRDS. Each violation entered into the system must contain the following:

o PWSID - The identification number of the system which violated the regulation
o Violation Type - The type of violation which occurred (maximum contaminant level, monitoring or reporting, etc.)
o Contaminant ID - An identifier for the contaminant whose regulation was violated

o Violation Date - The end of the compliance period for
   the particular violation

Additional information on the following may be available:

o Analytical Results - The analytical results of the analysis
o Analytical Method - A code indicating the analytical
   method used in the analysis

The analytical results and method are more likely to be found
for violations which occurred in community water systems, and
for violations of the inorganic, organic, or radiological
standards which occurred recently.

Contaminant Groups

The contaminants regulated by the NIPDWR can be divided
into four groups: microbials, inorganics, organics, and radionuclides. Following is an explanation of each group and the
primary limitations on the use of the data that are available
for community water systems.

Microbials

Total coliforms are regulated as the primary indicator of
the microbiological integrity of the water. The number of samples required varies by system size; results reported can be
either single sample results or system-wide average counts.
Turbidity is an indicator of the clarity of the water, and the
ability of the water to be effectively disinfected. Testing is
only required for surface water sources, and reported results
can either be single sample results or monthly average concentrations, measured at the entry point of the water to the water
distribution system.

Inorganics

These contaminants are measured yearly in surface water
systems and triennially in ground-water systems. Some are
routinely found in the environment: arsenic, barium, fluoride,
mercury, nitrate, selenium, and silver. Others are corrosion
by-products: cadmium, chromium, and lead. Since the monitoring
period for the contaminants, particularly in ground water, is
long, and the data are submitted to FRDS over a period of

years, multi-year scans of the data base must be conducted in order to prepare a comprehensive estimate of the occurrence of these contaminants.

Organics

Two groups of organic chemicals are regulated under the NIPDWR. The first is a group of six herbicides and pesticides (Endrin, Lindane, Methoxychlor, Toxaphene, 2,4-D, and 2,4,5-TP). Surface water systems monitor for these chemicals on a triennial basis. The second group, trihalomethanes, are controlled in systems which serve 10,000 or more individuals and which add a disinfectant to the water as part of the water treatment process. Trihalomethane monitoring is conducted quarterly, with compliance calculated on a rolling annual average of the quarterly concentrations. As with the inorganic chemicals, multi-year listings of violations are needed to prepare a comprehensive estimate of occurrence of these contaminants.

Radionuclides

Two groups of radionuclides are listed in the NIPDWR:

o Radium-226, Radium-228, and gross alpha
o Beta particle and photon radioactivity from manmade radionuclides (applied to community systems using surface water, serving more than 100,000 individuals, and designated by the state)

Again, these contaminants have multi-year monitoring periods, so that analysis of multi-year listings of reported violations is necessary to provide a complete assessment of occurrence.

CONCLUSION

Drinking water quality data are available from a number of sources. The data bases are all designed with special purposes in mind, and, unfortunately, research into the relationships between water quality and illness is not generally among the purposes. If research is to be conducted in this area, it is best to consider that the available data are related to contaminants that are regulated by EPA and the states. Data available from EPA relate to either currently regulated contaminants or to contaminants which are being considered for regulation.

These compliance data have been developed in the last ten years, include only exceedences of the Maximum Contaminants Limits, and are of little use in long-term studies relating water quality to disease incidence.

Information on water quality in state files will generally cover a longer time period than the data in EPA's system, but it is not likely to be available through an automated data retrieval system. State files will, however, be more likely to contain complete analytical results, not just exceedences recorded in the federal system.

Individual water system files will yield the most data over the longest time period, but the data are not likely to be automated or easy to obtain without spending time reviewing large volumes of written files.

The AWWA data base contains relatively recent data on the largest water systems in the country. Questions relating to specific information in the system of interest should be referred to AWWA, 6666 W. Quincy Ave., Denver, CO, 80235.

ACKNOWLEDGMENT

The paper from which this chapter is derived was developed to document the various sources of drinking water quality information. The work was performed as part of the routine program operation of the Office of Drinking Water at the U.S. Environmental Protection Agency.

CHAPTER 11

THE FDA TOTAL DIET STUDY PROGRAM

Pasquale Lombardo

INTRODUCTION

The Total Diet Study, also known as the Market Basket Study, is one of the U.S. Food and Drug Administration's (FDA) programs for monitoring chemical contaminants in foods. It is the only U.S. program that measures a broad range of these chemicals in foods as consumed. The principal objectives are to: (1) determine the dietary intake of pesticides, other industrial chemicals, elements (including heavy metals, radionuclides, and essential minerals); and (2) compare these intakes with Acceptable Daily Intakes (ADI), Recommended Dietary Allowances (RDA), or Estimated Safe and Adequate Daily Dietary Intakes. The program also allows identification of trends, may identify isolated contamination sources, and serves as a final check on the effectiveness of pertinent U.S. regulations and/or initiatives. The emphasis of this program is on pesticides.

BACKGROUND

The program was conceived in 1961, principally to determine whether the fallout from atmospheric nuclear tests resulted in elevated levels of radionuclides in foods. Analyses for pesticide residues were also part of the initial effort [1]. The foods examined comprised the "total diet" of a teenage male as based on data from the 1955 U.S. Department of Agriculture (USDA) Nationwide Food Consumption Survey [2] and the USDA Food Plan at Moderate Cost [3].

At the outset, "market baskets" containing a two-week supply of food (about 120 individual food items) were purchased at
the retail level in Washington, D.C., and prepared as for consumption (i.e., cooked or otherwise made table-ready). The prepared foods were separated into 12 groups of like foods (e.g.,
dairy products, leafy vegetables), and each group was blended
in amounts proportional to the weights of each in the diet of
the teenage male. Each food grouping (or composite) was then
analyzed; five FDA laboratories participated in the analyses of
the four market basket samples collected each year [4].

A number of modifications were made in subsequent years
[5-8]. These included: analyses for heavy metals and industrial chemicals; modifications of the teenage diet to reflect
more recent food consumption data; increasing the number of annual market basket collections to 30; collection of the market
baskets in different cities across the country; centralizing
the analyses in the FDA Kansas City District in 1970; analyses
for nutrient elements in 1974; and inclusion of separate market
baskets for infants and toddlers in 1975. At that point, the
annual collections comprised 20 teenage and 10 infant-toddler
baskets. The food groupings analyzed for each population group
are shown in Table 1.

Table 1. Food Groupings.

Teenage Diet	Infant Toddler Diet
Dairy products	Drinking water
Meat, fish & poultry	Whole milk
Grain & cereal products	Other dairy & dairy
Potatoes	substitutes
Leafy vegetables	Meat, fish & poultry
Legume vegetables	Grain & cereal products
Root vegetables	Potatoes
Garden fruits	Vegetables
Fruits	Fruits & fruit juices
Oils & fats	Oils & fats
Sugar & adjuncts	Sugar & adjuncts
Beverages	Beverages

The foods represented a typical 14-day diet or subset of
the total food supply. The contaminant and mineral content of
each food subset was extrapolated in proportion to the weight
consumed to allow estimation of the daily contaminant and mineral intakes of the three age-sex groups (6 month old infants,
2 year old toddlers, and 16-19 year old males).

CURRENT STUDY

The most significant change took place in 1982; after two years of intensive planning, the Total Diet Study was completely redesigned. Selection of the diets was based on two nationwide surveys covering about 50,000 people: the 1977-78 USDA Nationwide Food Consumption Survey [9] and the 1976-80 Second National Health and Nutrition Examination Survey [10]. About 5000 different foods were identified in these surveys.

Practical considerations precluded the collection and analysis of the approximately 900 foods required to represent 95% by weight of the average diet, or even the 500 foods required for 90% representation [11]. Using an aggregation scheme, 234 foods were selected to represent the 5000 foods [11,12]. For example, "apple pie" represents dozens of different fruit pies and pastry with fruit, "beef and vegetable stew" represents mixed dishes which contain meat with potato or other starchy vegetables plus other vegetables in a gravy or sauce, and "chocolate milkshake" represents all types of malts, milkshakes, and eggnogs. The same surrogate foods are always chosen. No brand names are specified, thus the selection is random. These 234 foods can be said to represent all the foods eaten in this country. The former composite approach was terminated in favor of chemical analyses of each of the 234 foods. Analysis of individual foods enabled the construction of diets for eight age-sex groups (6 to 11 month olds, 2 year olds, and 14-16, 25-30, and 60-65 year old males and females) as compared to the previous three. Additionally, the elimination of the "dilution effect" inherent in the composite approach enables the detection of analytes that would previously have gone unnoticed.

Under the present scheme, the food items are purchased at retail stores in each of four geographic areas (northeast, north central, south, and west) to give a total of four market baskets per year. Each basket is composed of foods collected simultaneously in three cities in one of the geographic areas. Collections by geographic areas are rotated, e.g., the northeast collection may take place in the spring of one year and the fall of the next. The cities within each geographic area are changed with each collection. The foods are shipped to the Kansas City Total Diet Laboratory, where the three samples of each particular food item are combined and prepared as for consumption. Each of the 234 prepared foods is then analyzed individually for residues of over 100 pesticides, many industrial chemicals (such as polychlorinated biphenyls [PCB]), heavy metals and essential minerals (Cd, Pb, As, Se, Hg, Zn, Cu, Fe, Mg, Mn, K, P, Ca, Na, and I), and radionuclides (^{90}Sr, ^{137}Cs, ^{131}I, ^{106}Ru, and ^{40}K). Most of the analyses are carried out using multi-analyte analytical methods; five different methods are used for the pesticides [13]. Dietary intakes are then calculated for the eight age-sex groups.

The program is conducted to determine levels of chemical contaminants in foods as eaten rather than to enforce tolerances or other regulatory limits for residues on raw agricultural commodities. The analytical procedures have been modified to permit quantification at levels five to ten times lower than those attained in FDA regulatory monitoring programs (a greater equivalent sample weight is presented to the determinative step). The identity of each organic chemical reported is confirmed by an alternative method and frequent blank and recovery analyses are conducted on a variety of food/analyte combinations to monitor and ensure acceptable analytical method performance.

DISCUSSION

Typically, 80-90 different chemicals are found in each current market basket. Of the more than 200 pesticides and associated chemicals that are detectable, about 60 are present in each basket. Malathion, a widely used insecticide, and DDE, a metabolite of DDT, have been the most frequently found pesticide residues. In general, the residue levels of pesticides are much lower than those specified in the tolerances and their calculated dietary intakes fall well below established ADIs. About a dozen industrial chemicals are usually found, and as expected, there are many findings of the essential minerals and some of the heavy metals. It is reassuring to note that the radionuclide levels remain very low or at "background;" this has been the case since the early years of the program.

Trends may be identified. The dietary intakes of many persistent chlorinated pesticides have steadily declined since the chemicals were banned ten or more years ago. For example, the calculated intake of dieldrin approached the ADI in the late 1960s (the only chemical to have done so); present-day intake is only a small fraction of the ADI. DDT intake has dropped dramatically since its uses were cancelled. Because of its environmental persistence, DDT residues (chiefly in the form of DDE) continue to be found in many foods, albeit at low levels. The decline in PCB intake is also notable; in several of the more recent market baskets, none of the 234 food items contained detectable PCB residues. The intakes of the heavy metals have remained relatively constant or have dropped only slowly over the years. For lead, though, the analysis of individual foods has permitted following the decline of its levels in canned foods; the concentrations are about one tenth of what they were about ten years ago. This reduction can be attributed to industry's continuing conversion to nonlead-soldered cans as well as overall improvements in the manufacture of soldered cans. Table 2 lists average daily dietary intakes of several chemical contaminants over the past 20 years.

Table 2.   Average Daily Intake (μ g) of Selected Contaminants.[a]

Contaminant	FAO/ WHO ADI[b]	1965- 1970	1971- 1976	1977- 1982	1982- 1984
DDT (total)	300	31.0	4.5	2.5	2.5
Dieldrin	6	3.1	1.8	0.8	0.4
Endrin	12	0.3	0.1	0.01	0.01
Heptachlor Epoxide	30	1.4	0.3	0.3	0.2
PCB	None	c	1.4	0.6	0.03
Cadmium	57-72[d]	24.5[e]	23.3	20.0	15.4
Lead	429[d]	c	46.5[f]	51.0	41.3
Mercury	43[d]	c	2.6	2.6	2.6
Strontium-90	0-20[g]	17.8	7.8[h]	6.5[i]	4.9

[a]Teenage male, basis 2520 kcal/day diet.
[b]ADI converted from mg/kg body weight/day to μg/day, basis
 60 kg body weight.
[c]Not analyzed during this time period.
[d]FAO/WHO Provisional Tolerable Weekly Intakes converted to
 daily intakes for purposes of comparison.
[e]Three-year average (1968-1970).
[f]Four-year average (1973-1976).
[g]Federal Radiation Council intake range in pCi/day, for which
 only periodic surveillance is recommended.
[h]Three-year average (1974-1976), pCi/day.
[i]Four-year average (1977-1980), pCi/day.

    Occasionally, unexpected findings surface. About 15 years
ago, PCB residues were found in a dry cereal. Follow-up investigation revealed that the chemical had migrated from the card-
board package made from PCB-contaminated recycled paper. This
finding ultimately led to regulations limiting the PCB content
of paperboard intended for food-contact use. In another instance, a residue of the preservative/fungicide pentachloro-
phenol (PCP) was found in unflavored gelatin. It was later
learned that past uses of PCP included treatment of hides in
slaughterhouses to inhibit spoilage during storage and that
many of these hides were sent to gelatin manufacturers. This
use of PCP had been discontinued by the United States industry
several years prior to the finding, and investigation at the
gelatin manufacturing facility revealed that the sample in
question was a mixture of domestic and Mexican gelatin. The
Mexican gelatin was found to contain the PCP and was ultimately
diverted from food use. In the nutrient area, calculated
iodine intakes were found to be several-fold higher than the
RDA. For adults, major contributors of iodine were dairy products, grains and cereal products, meat, fish, and poultry, and
sugars and adjuncts. For infants and toddlers, milk, other

dairy products, and grains and cereals were major contributors. The information helped FDA identify specific problem commodities, and appropriate segments of industry were advised in efforts to encourage voluntary reduction in iodine usage. Regulatory limits were also established in some cases.

The information generated by the program has many uses: it is in constant demand by Congress, industry, the news media, consumer groups, national and international organizations, and government agencies.   In particular, the "real world" dietary exposure data play an important role in EPA's continuing reassessment of pesticide tolerances.   Perhaps the most important user of the data is FDA itself, as the information serves to guide or redirect many of its monitoring, regulatory, and research activities.   As mentioned earlier, the data also provide a final check on the effectiveness of the United States regulatory system for pesticides.   The contaminants information is currently being published in the Journal of the Association of Official Analytical Chemists [14,15] and the essential minerals data appear in the Journal of the American Dietetic Association [16].

As an integral part of FDA's overall program on pesticide residues, the Total Diet Study complements the agency's other monitoring activities, which focus primarily on the raw agricultural commodity.   The study is cost-effective, as it is much more resource-intensive to carry out ad hoc, nonsystematic analyses of many different foods to develop equivalent information on the wide spectrum of chemicals covered.   The program also helps promote consumer confidence in the safety of the food supply, since the chemical residues measured in foods as eaten demonstrate low dietary intakes of contaminants.   Thus, the public may be assured that the food supply does not contain excessive residues of "poisons."   The negative findings are of equal importance because they indicate the absence of many chemicals in the food supply.

The program is continuously evolving and ways are currently being explored to expand the coverage without increasing resources.   Finally, it is the only United States program that measures a broad range of chemicals in foods as consumed. Thus, with empirical data in hand, FDA does not have to rely on theoretical estimates or "best guesses."   The FDA Total Diet Study has also served as a model for many countries throughout the world; this peer acceptance may be taken as a measure of its success.

There are, however, some limitations.   The program does not cover all analytes of interest.   For example, less than half of the 300 or so registered pesticides are determined.   The program, though, covers most of the important ones, i.e., the environmentally persistent chemicals that can biomagnify through the food chain and produce chronic toxic effects.   Only a nationwide picture is developed; the study does not provide information on special populations or ethnic groups.   Only four data points (one per market basket) are developed each year; several years are usually needed before trends become evident or conclusions reached. Also, logistical considerations generally prevent ad hoc insertion of new analytes into the program, as the net effect might well be too disruptive. Finally, "Total Diet Study" is a title that may imply something the program is not; it does not measure nutritional quality, adequacy of the American diet, or food intake, as the title alone may indicate.

In sum, the Total Diet Study helps to fulfull FDA's responsibility to determine the incidence and level of contaminants and selected nutrient minerals, and helps promote consumer confidence in the safety of the food supply. It is the only program of its kind in the United States, has been universally recognized, and is an effective means to measure dietary intakes of a host of contaminants and nutrients. Finally, the program continues to provide a measure of the effectiveness of United States regulations and initiatives on pesticides, chemical contaminants, and selected nutrients.

DISCLAIMER

The work described in this chapter has not been funded by the EPA and no official endorsement should be inferred.

REFERENCES

1. Laug, E. P., A. Mikalis, H. M. Bollinger, and J. M. Dimitroff. "Total Diet Study," *J. Assoc. Off. Agric. Chem.* 46:749-767 (1963).

2. "Food Consumption of Households in the United States, Household Food Consumption Survey, 1955, Report 1," U.S. Dept. of Agric. (1956).

3. "Family Food Plans and Food Costs," Home Economics Research Report #20, Agricultural Research Service, U.S. Dept. Agric. (1962).

4. Williams, S. "Pesticide Residues in Total Diet Samples," *J. Assoc. Off. Agric. Chem.* 47:815-821 (1964).

5. Duggan, R. E., and F. J. McFarland. "Residues in Food and Feed," *Pestic. Monit. J.* 1:1-5 (1967).

6. Duggan, R. E., and H. R. Cook. "National Food and Feed Monitoring Program," *Pestic. Monit. J.* 5:37-43 (1971).

7. Manske, D. D., and P. E. Corneliussen. "Pesticide Residues in Total Diet Samples (VII)," Pestic. Monit. J. 8:110-124 (1974).

8. Johnson, R. D., D. D. Manske, D. H. New, and D. S. Podrebarac. "Pesticides and Other Chemical Residues in Infant and Toddler Diet Samples - (I) - August 1974-July 1975," Pestic. Monit. J. 13:87-98 (1979).

9. "U.S. Dept. Agric. Nationwide Food Consumption Survey, Spring, Summer, Fall and Winter Quarters, 1977-78," National Technical Information Service, Springfield, VA, Accession Numbers PB 80-190218, PB 80-197429, PB 80-200223 and PB 81-118853.

10. "Second National Health and Nutrition Examination Survey, 1976-80," National Technical Information Service, Springfield, VA, Accession Number PB 82-142639.

11. Pennington, J. A. T. "Revision of the Total Diet Study Food List and Diets," J. Am. Dietet. Assoc. 82:166-173 (1983).

12. "Documentation for the Revised Total Diet Study: Food List and Diets," National Technical Information Service, Springfield, VA, Accession Number PB 82-192154.

13. "Pesticide Analytical Manual," Food and Drug Administration, Washington, DC (1968 and revisions), Vol. I, Secs. 211.1, 212.1, 221, 231.1, 232.1, and Appendix.

14. Gartrell, M. J., J. C. Craun, D. S. Podrebarac, and E. L. Gunderson. "Pesticides, Selected Elements, and Other Chemicals in Infant and Toddler Total Diet Samples, October 1980-March 1982," J. Assoc. Off. Anal. Chem. 69:123-145 (1986).

15. Gartrell, M. J., J. C. Craun, D. S. Podrebarac, and E. L. Gunderson. "Pesticides, Selected Elements, and Other Chemicals in Adult Total Diet Samples, October 1980-March 1982," J. Assoc. Off. Anal. Chem. 69:146-161 (1986).

16. Pennington, J. A. T., D. B. Wilson, R. F. Newell, B. F. Harland, R. D. Johnson, and J. E. Vanderveen. "Selected Minerals in Foods Surveys, 1974 to 1981/82," J. Am. Dietet. Assoc. 84:771-782 (1984).

CHAPTER 12

OVERVIEW OF EPA MAJOR AIR DATA BASES

David W. Armentrout

INTRODUCTION

This chapter provides a brief overview of the content and
capabilities of the primary air data bases maintained by the
U.S. Environmental Protection Agency (EPA).

The EPA maintains several air data bases, which include
emissions-related data and ambient air quality monitoring data. These data bases concentrate primarily on the criteria air
pollutants, i.e., those for which national ambient air quality
standards have been adopted (total suspended particulates, sulfur dioxide, nitrogen dioxide, carbon monoxide, ozone, and
lead). They also include some ambient monitoring data for selected hazardous air pollutants, but these data are not extensive. The criteria pollutant data are used by EPA primarily to
track ambient air quality and, through dispersion modeling
techniques, to evaluate air quality control strategies and environmental policy options.

The National Air Data Branch (NADB) of the Office of Air
Quality Planning and Standards maintains the primary air data
bases at the National Computer Center at Research Triangle
Park, North Carolina. The comprehensive system of data bases
is called the Aerometric and Emissions Reporting System
(AEROS). The primary subsystems originally included the following:

o National Emissions Data System (NEDS) - Source-specific
   emissions data including stack parameters and operating
   rates for major emitting facilities.
o Storage and Retrieval of Aerometric Data System (SAROAD)
   - Ambient air monitoring data from monitoring sites located nationwide.

o Hazardous and Trace Emissions System (HATREMS) - Emissions data for selected hazardous pollutants.
o Source Test Data System (SOTDAT) - Selected data from stack emissions testing.
o Quality Assurance Management Information System (QAMIS) - Data on quality assurance for specific air monitoring sites.

Only the NEDS and SAROAD systems have been developed and used to any important extent. The NEDS and SAROAD systems include extensive data bases and sophisticated storage and retrieval capabilities. These systems have evolved through years of analysis of the needs of air regulatory agencies for access to the data. They are based on mandatory data submittals from state and local regulatory agencies as mandated by the Clean Air Act. These two systems provide input to the tracking of the effectiveness of air quality control programs and to the development, refinement, and assessment of regulatory control strategies.

Because the data in NEDS and SAROAD are critical to strategy development and program assessment, the data are submitted to a series of validation checks prior to being entered into the data bases. Data are submitted by the state and local agencies to the EPA regional offices where they are subjected to system validation features. Questionable and incomplete data are returned to the submitting agency for problem resolution before they are entered into the data bases. The data are updated periodically according to a set schedule (quarterly for air quality data and annually for emissions data).

To assist state agencies in meeting their submittal requirements and in implementing in-house data base capabilities, NADB developed a data system patterned after NEDS and SAROAD for implementation at the state level. This system, called the Comprehensive Data Handling System (CDHS), includes software maintained by NADB. It allows the state to meet formal EPA reporting requirements through tape submittal, as opposed to the traditional method of hard copy data submittal. The states have the flexibility of maintaining data useful to their specific regulatory programs as well as data required by EPA.

Development of the NEDS data base began in the early 1970s with EPA funding contractor efforts to code pertinent data on major emissions sources from agency permit files. Updates now consist of data entered for new sources or modifications at existing sources. The data base represents approximately 36,000 facilities. An estimated 11,000 of these facilities each have emissions greater than 100 tons per year, and approximately 3000 facilities each have emissions greater than 1000 tons per year.

The facility-specific data in NEDS include source information necessary to characterize individual emission points within each facility with respect to location, stack parameters and emission control equipment, type of combustion source or process, estimated annual emissions, and emissions rates allowed by regulation. For example, a facility may have several emission points. The data for each emission point would include:

o Point identification number
o Year of record
o Standard Industrial Classification (SIC) code
o UTM coordinates
o Stack data (height, diameter, temperature, flow rate)
o Indicator of processes which emit through the same stack
o Boiler capacity (if applicable)
o Control equipment and rated efficiency for each pollutant
o Quarterly percent throughput
o Emissions estimates and method used to estimate
o Allowable emissions
o Source Classification Code (SCC)
o Annual fuel consumption or process operating rate
o Maximum process design rate

Historical records are not maintained. The data for each point source represent a snapshot of the source for the year indicated in the record.

Point source data may be retrieved for individual point sources and facilities or for multiple point sources within a selection category. For example, a retrieval could include the point source data for all sources within a state, an Air Quality Control Region (AQCR), or a specific facility. Point source data also could be retrieved by ownership code (public or private facility), SIC, emissions estimate method, SCC code, emissions volume classification, or any combination of these and/or the geographic retrieval codes. The SCC identifies the specific type of process or combustion unit represented by each NEDS record. This code is particularly useful in retrieving data for similar source types at different facilities. For example, data for all coal-fired boilers of a specified size could be retrieved by keying on a single SCC. AEROS contact personnel within each EPA regional office can provide information on specific classification codes.

The NEDS system also provides emissions summaries by pollutant for specified retrieval parameters. For example, emissions can be shown for all of the NEDS pollutants for combustion source and industrial process source categories within a specified geographical area.

The SAROAD data base includes historical data on both previously active and currently active ambient air monitoring sites. Data in SAROAD, unlike NEDS data, represent historical records, and may date back prior to 1970 for some monitoring sites. These include sites operated by state and local agencies for their own programs, sites operated by private businesses, and sites operated by the state and local agencies for EPA. The sites operated for EPA are designated as National Air Monitoring Sites (NAMS). These are fixed sites established specifically to provide data for studying air quality trends. These sites were screened with respect to EPA siting criteria to eliminate or minimize bias from specific emissions sources.

Further, they were selected based on specific varying population and industrial concentration and geographic area representation. These sites were screened for conformity to EPA guidelines for monitor siting which stipulate siting parameters such
as setback from roadways (e.g., for lead samples) and height
above ground. EPA maintains site descriptions for these sites
for use in data interpretation. The initial screening of the
NAMS sites occurred in 1977.

The SAROAD data base contains identifier and locator data
for each monitoring site, including supporting agency, city and
county, site address, latitude/longitude, UTM coordinates, and
elevation (above ground and above sea level). The raw data
records for each site include:

o Site identification
o Parameter observed (may include meteorological parameters)
o Parameter code
o Time interval
o Monitoring method
o Reporting units
o Data values based on appropriate monitoring or averaging
   times for each pollutant

Raw data reports show site descriptor information and individual parameter values. These reports also show the number of
observations, average values, and maximum values for the reporting periods.

Summary and management reports are also available. Annual
and quarterly frequency distributions, for example, can be retrieved for each site. Management reports include site inventory by geographical area, site inventory by pollutant, and inventory of active sites. Retrievals are available by EPA region, state, site, AQCR, year, pollutant, and other retrieval
designators or combinations of designators, depending on the
type of report being requested.

NADB currently is in the planning phase of a major effort
to replace the software for NEDS and SAROAD. The new system
will incorporate state-of-the-art data management capabilities
and expanded data analysis capabilities. The implementation
effort is estimated as an approximately three-year effort.

The new system development includes plans for an update and
data confirmation of NEDS data for major emissions sources,
i.e., sources emitting greater than 100 tons per year of any
criteria pollutant. The new data base also would incorporate
selected compliance tracking data. Currently, NEDS does not
include compliance tracking data. These data are maintained in
a separate Compliance Data System (CDS) maintained by the Stationary Source Compliance Division. Combining data from these
two systems will be a major coordination effort, since individual emission source information is often difficult to cross-
reference between the data bases.

The restructuring of the air data systems should allow for
the incorporation of data on toxic air pollutants. These data

currently are not included, and there is no regulatory requirement for data collection and reporting. The current system
does, however, include codes for a variety of toxic pollutants.

The EPA is involved in a significant effort to characterize
sources of toxic emissions and to develop methods for ambient
monitoring of toxics. A research and development effort is underway to develop methods for toxic pollutant monitoring. The
current effort includes a system of Toxic Air Monitoring Sites
(TAMS) established in Boston, Chicago, and Houston. It is not
known when or to what extent EPA will regulate and require
emissions data and ambient air monitoring data to be reported
for toxic pollutants. It is anticipated that current NADB system development efforts will incorporate a capability for
toxics data to be included in the data bases. It probably will
be several years, however, before extensive toxics data are included in the data bases.

DISCLAIMER

The work described in this chapter was not funded by EPA
and no official endorsement should be inferred.

CHAPTER 13

NATIONAL DATABASE ON BODY BURDEN OF TOXIC CHEMICALS

Philip E. Robinson, Cindy R. Stroup, M. Virginia Cone,
Marialice Ferguson, Anna S. Hammons, C. Donald Powers,
and Herman Kraybill

INTRODUCTION

The National Database on Body Burden of Toxic Chemicals is
composed of two major files, Chemicals Identified in Human Biological Media and Chemicals Identified in Feral and Food Animals, which were established in 1978 and 1980, respectively,
under the aegis of the Interagency Collaborative Group on Environmental Carcinogenesis, National Cancer Institute (NCI). The
program to develop and maintain the data base is funded through
the NCI/Environmental Protection Agency (EPA) Collaborative
Program and an interagency agreement between the EPA and the
Department of Energy (DOE). The work is conducted under the
management of the Office of Toxic Substances (OTS), EPA, by
Science Applications International Corporation (SAIC).

The concept of a national resource for body-burden data developed from concerns of the scientific community over continuing reports of toxic chemicals being found in human tissues and
fluids. Scientists recognized the necessity for a comprehensive, centralized, and available source of data concerning human body burden of xenobiotics. Such data are needed to assist
in identifying industrial chemicals of concern and in setting
priorities for nomination and selection of chemicals for carcinogenesis bioassay.

SOURCES, FORMATTING, AND DISTRIBUTION OF DATA

Data for this program are from the world literature, retrospective to 1974 and 1978 for the human and animal files, respectively. Approximately 60 periodicals are routinely searched manually for current literature. Also, contacts with federal agencies and private investigators have been made to identify pertinent body-burden data for inclusion in the data base. The data base currently contains information on about 1300 chemicals.

The companion animal file was established to complement the human file because 1) animals provide a significant portion of the human food chain, 2) animal body burdens of environmentally ubiquitous chemicals provide an early warning of potential human exposure, and 3) various species of animals are better indicators of exposure than humans because observable health and physiological effects occur at much lower concentrations than in humans. About 45 periodicals are searched routinely for data concerning animals.

Each record in the data base contains information on a specific chemical/tissue or chemical/tissue/animal combination. Thus, a single source document may contain material that yields multiple records. Specific elements in the data base are listed in Table 1.

Table 1.   Elements in Data Base.

Analytical technique	Half-life
Animal, species	Key words
Bibliographic information	Language (other than English)
CAS Registry number	Levels measured (mean, range)
Chemical Abstracts Service (CAS)	Number of cases
preferred name	Organ, tissue, or body fluid
Chemical formula, properties	Pathology, morphology
Data source (report, journal,	Route of exposure
letter)	Source of chemical
Demography	Synonyms
Explanatory comments or caveats	Toxicity

New records, arranged alphabetically by CAS preferred name and in tabular format, are published annually. Author, corporate author, tissue, and keyword indices are included, as well as several appendices and a directory of chemicals. These publications are distributed internationally to libraries of government agencies, medical schools, public health institutions, and to various universities. The following publications are available from the National Technical Information Service,

Springfield, VA 22161: Chemicals Identified in Human Biological Media, A Data Base. Volumes I-VII for 1979-1984 may be ordered by the numbers ORNL/EIS-163/V1-V6, and EPA-560/5-84-003, Chemicals Identified in Feral and Food Animals, A Data Base. Volumes I-IV for 1981-1984 may be ordered by the numbers ORNL/EIS-196/V1-V3, and EPA-560/5-84-004. For access to all records on-line (File 138), write to DIALOG$^{(R)}$, Information Retrieval Service, 3460 Hillview Avenue, Palo Alto, CA 94304. Currently, plans are being developed to also make the data base available through the National Library of Medicine.

USES OF THE DATA BASE

Surveys of the users show that the data base is especially important to those involved in assessments of risk associated with exposure to toxicants, in toxicological research and testing, and in disease prevention and treatment.

Body-burden data on toxic chemicals provide "de facto" evidence that exposure has occurred. Such information is important to OTS because it is sufficient to require that toxicological testing be performed when such data are unavailable. Also, priority setting, historically done on a toxicological basis, can now be performed by focusing initially on those chemicals for which exposure has occurred.

The availability of an organized, comprehensive, body-burden data base facilitates the early identification of human exposure to environmental contaminants and aids in assessing the significance of such exposure. The OTS is currently using this data base to help identify chemicals in the Chemical Substances Inventory, mandated by the Toxic Substances Control Act, that pose a potential risk to the general population. Subsequent actions to be taken might include the requirement for further toxicological testing if available data are not sufficient for an assessment, or placement of the chemical in the Existing Chemical Assessment Process for a more detailed evaluation of the exposures and risks posed by the chemical.

Additionally, the data base can help identify populations at increased risk as well as probable sources of exposure. This information, while not sufficient for regulatory purposes, does provide important information around which to plan further activities directed at developing statistically valid estimates of exposure and risk.

The data base is used by medical and health professionals in teaching and by investigators in planning research in toxicology, epidemiology, and monitoring. Body-burden data often serve as a baseline against which comparisons can be made, using data collected in research on the levels and frequency of detection of selected chemicals in human tissues and fluids.

FUTURE DEVELOPMENT

An evaluation of the data base is currently underway. The completed Phase I of this evaluation was aimed at assessing the utility of the data base. The highly favorable response to our user survey showed that the data base is used extensively by regulators, researchers, and other health and environmental professionals and students, and that users are generally pleased with the format and content. Nevertheless, some changes may prove helpful. Phase II of the evaluation addresses modifications to content and format that would facilitate use as well as increase comprehensiveness. The following changes are either being considered or are being implemented in the next annual report:

o Adding a cumulative index of chemicals.
o Dividing drugs and non-drugs into separate volumes.
o Including data on edible plants.
o Classifying animals as vertebrate/invertebrate and domestic/wild.
o Grouping chemicals by class.
o Updating the on-line systems quarterly.

As a long-range goal, the development of appropriate mechanisms for computerized scanning would ultimately provide the most efficient and cost-effective way to collect body-burden data from the open literature. As those in the business of publishing these data become aware of the need to facilitate the identification of such information, such techniques will be developed and implemented.

CONCLUSIONS

The results of the Phase I evaluation have verified the extensive need for the body-burden data base, particularly by regulators, researchers, and other health-oriented professionals, students, and government agencies. While it is impossible to accurately identify either all of the users or the uses, the approximately 250 responders to our user survey indicate that the data base is valuable to the following major users:

o Technical experts in government agencies in nominating and selecting chemicals for various bioassays and in performing various assessments of specific chemicals.
o Medical and public health professionals in teaching and in assessing risk to exposed individuals or populations.
o Researchers in planning research and comparing research results.
o Data base managers to augment other data bases.

DATABASE ON BODY BURDEN OF TOXIC CHEMICALS   159

Several changes designed to make the annual reports easier to use have recently been implemented and will appear in the next annual publication. For example, a cumulative index of chemicals, recommended by several of our users, will be included in future reports.

Users' responses to the data base have been extremely positive and constructive. Recommended changes to enhance the utility of this resource are always welcome and are carefully considered. Improvements to the data base are ongoing, and we have found that the best way to identify and implement improvements is by encouraging continuing dialogue between the users and the data base managers.

Data base activities are focused not only on providing a comprehensive, national resource for body-burden data, but also on working with the users to ensure that this resource can be readily accessed and is easy to use. Future plans include further development of the program to provide other products, such as specialized summary reports and bibliographies.

ACKNOWLEDGMENTS

This program is funded through the National Cancer Institute/Environmental Protection Agency Collaborative Program and an interagency agreement between the Environmental Protection Agency and the Department of Energy (EPA No. DW89930139-01-4, DOE No. 40-822-84).

CHAPTER 14

BROAD SCAN ANALYSIS OF HUMAN ADIPOSE TISSUE
FROM THE EPA FY 82 NHATS REPOSITORY

John S. Stanley, Kathy E. Boggess, John E. Going,
Gregory A. Mack, Janet C. Remmers, Joseph J. Breen,
Frederick W. Kutz, Joseph Carra, and Philip Robinson

INTRODUCTION

The U.S. Environmental Protection Agency's Office of Toxic
Substances (OTS) maintains a unique capability for estimating
exposure of the general United States population to toxic organic chemicals. The National Human Adipose Tissue Survey
(NHATS) is the main operative program of the National Human
Monitoring Program (NHMP), which is an ongoing chemical monitoring network designed to fulfill the human monitoring mandates of the Toxic Substances Control Act (TSCA). The NHMP was
first established by the U.S. Public Health Service in 1967,
and was transferred to EPA in 1970. In 1979 the program was
transferred within EPA to the Exposure Evaluation Division of
OTS.

NHATS is an annual program whose purpose is to collect and
chemically analyze a nationwide sample of adipose tissue specimens for the presence of toxic substances. The objective of
the NHATS program is to detect and quantify the prevalences of
selected toxic compounds in the general population, which historically have been organochlorine pesticides and polychlorinated biphenyls (PCB) [1-6]. The specimens are collected from
surgical patients and autopsied cadavers according to a statistical survey design. The survey design ensures that specific
geographic regions and demographic categories are appropriately
represented to permit valid and precise estimates of baseline
levels, time trends, and comparisons across subpopulations.

EPA/OTS has developed an aggressive strategy to expand the
use of the NHATS specimens to provide a more comprehensive assessment of TSCA-related substances that are persistent in the

human adipose tissues of the general United States population.
The NHATS specimens collected during fiscal year (FY) 1982 were
selected for broad scan analysis of volatile and semivolatile
organic chemicals and trace elements.

This initiative for a more comprehensive assessment of
toxic subtances in human adipose tissue necessitated either the
development of new methods or the modification of existing procedures. Data reported on NHATS specimens up to the FY 82 collection have been focused on organochlorine pesticides and PCB,
based on packed column gas chromatography/electron capture detector (PGC/ECD) analysis. However, preliminary data for polychlorinated terphenyls and polybrominated biphenyls from the
gas chromatography/mass spectrometry (GC-MS) analysis of pooled
NHATS specimen extracts from previous collection years have
been reported [7,8].

The objectives of the broad scan analysis program were to:
identify appropriate analytical methods based on high resolution gas chromatography/mass spectrometry (HRGC/MS) detection
for general semivolatile and volatile organic compounds and on
two multielemental techniques - neutron activation analysis
(NAA) and inductively coupled emission spectrometry (ICP-AES) -
for toxic trace elements; conduct preliminary evaluation of the
analytical procedures; complete the sample workup and HRGC/MS
analysis of 46 composite samples prepared from over 750 NHATS
specimens collected during FY 82; and compare the data generated by the two multielemental techniques through the analysis
of nine individual NHATS specimens.

The broad scan analysis approach based on HRGC/MS and the
multielemental techniques were necessary to identify additional
compounds or toxic trace elements that might be of concern to
EPA under the mandates of TSCA. The multielemental analysis
techniques were included as screening procedures to provide information on toxic trace elements that persist in human adipose
tissue. The analytical procedures used and the results that
were generated are summarized here.

EXPERIMENTAL

Sample Collection

A nationwide random sample of selected pathologists and
medical examiners collect and send to EPA/OTS adipose tissue
specimens extracted from surgical patients and cadavers on a
continuing basis throughout each fiscal year. In order to develop statistically valid information on a national basis, collections of adipose tissue are achieved according to a survey

design that dictates the number of samples required. Sample quotas reflect the demographic distribution of the population in the specific census divisions. The cooperating pathologists and medical examiners are provided the necessary sampling supplies (i.e., chemically clean specimen bottles, specimen labels, shipping materials, etc.), criteria for collecting samples, and instructions or methods to reduce potential background contamination of individual specimens. The specimens are frozen (-20° C) immediately after collection and transferred to the NHATS repository. The pathologists and medical examiners supply EPA with a limited amount of demographic, occupational, and medical information with each specimen. This information allows reporting of residue levels by subpopulations of interest, namely by sex, race, age, and geographic region.

Compositing Scheme

Composite samples of approximately 20 g each were prepared from more than 750 specimens from the NHATS FY 82 repository. The compositing scheme resulted in samples representing the nine U.S. Census divisions and three age groups (0-14, 15-44, and 45+ years). Additional composites of particular age groups within a census division were prepared to demonstrate variability in preparing composites and variability based on sex or race (white/nonwhite). The composites were prepared by weighing and combining 1.0- to 2.0-g aliquots of each specimen identified in the sampling design. Composite samples were prepared for both the semivolatile and volatile organic compound analytical procedures. All samples were handled in a positive pressure Plexiglas™ hood of approximately 94.5 l volume to prevent contamination from laboratory air. Compressed air was filtered through a charcoal trap before entering the hood. The individual samples were manipulated with stainless steel spatulas and placed in glass vials and sealed with Teflon™ septa caps. The composited samples in the sealed vials were placed in 1 qt jars containing a layer of activated charcoal and sealed with a Teflon™-lined lid. All composites were stored at -20° C until analysis. Blanks were included with the composites and consisted of empty glass vials taken through the same cleanup and laboratory conditions as the actual composited samples. The composites prepared for analysis of general semivolatile organic compounds were also used for specific analyses for toxaphene and polychlorinated dibenzo-p-dioxins (PCDD) and polychlorinated dibenzofurans (PCDF).

Analytical Procedures

Semivolatile Organic Compounds

Figure 1 provides a schematic of the method followed for the broad scan analysis of semivolatile organic compounds. Several stable isotope labeled compounds were added to the composited tissue as surrogate analytes. The surrogates included:

naphthalene-d_8 (2 μg)
chrysene-d_{12} (2 μg)
1,2,4,5- tetrachlorobenzene-$^{13}C_6$ (2 μg)
4-chlorobiphenyl-$^{13}C_6$ (2 μg)
3,3'4,4'-tetrachlorobiphenyl-$^{13}C_{12}$ (4 μg)
2,2',3,3',5,5',6,6'-octachlorobiphenyl-$^{13}C_{12}$ (8 μg)
decachlorobiphenyl-$^{13}C_{12}$ (10 μg)
2,3,7,8-tetrachlorobenzo- p-dioxin-$^{13}C_{12}$ (1 ng)
and octachlorodibenzo- p-dioxin-$^{13}C_{12}$ (5 ng).

The spiked adipose tissue sample was extracted with five 10-ml aliquots of methylene chloride using a Tekmar Tissue-mizer™. The extracts were filtered through anhydrous sodium sulfate and the final volume was adjusted to 100 ml. Extractable lipid was determined gravimetrically using approximately 1% of the resulting extract. The extracts were concentrated to achieve approximately 0.3 g lipid/ml, and the lipid was separated from organic analytes using gel permeation chromatography (GPC).

The GPC columns were prepared with 60 g of Bio Beads™ SX-3 (BioRad Laboratories) swelled in methylene chloride and packed as a slurry. The GPC was operated with methylene chloride as the mobile phase at 5 ml/min under a pressure of 7-15 psi. Typical GPC operating conditions were: sample size 0.9-1.0 g lipid per sample loop, discard the first 25 min of eluent containing lipids and collect eluent from 25-60 min. Total cleaning time per sample loop was approximately 60 min.

The GPC-cleaned extracts were concentrated using Kuderna Danish evaporators and then fractionated using Florisil™ [2]. The semivolatile organic compounds were eluted from the Florisil™ using 6%, 15%, and 50% diethyl ether/hexane solvent mixtures. These Florisil fractions were exchanged to hexane using Kuderna Danish evaporators and concentrated to 200 l using flowing purified nitrogen, spiked with an internal quantification standard (anthracene-d_{10}, 2 μg), and analyzed by HRGC/MS.

Separation of analytes was achieved using a 30 meter x 025 mm Durabond™ DB-5 0.25 μm film thickness, HRGC column. Sample extracts were injected through a Grob style splitless detector. The HRGC column was held isothermal for 2 min, then programmed at 10° C/min to a final temperature of 310° C. The ion source of a Finnigan MAT 311A double focusing magnetic sector mass spectrometer was operated at 70 eV. A mass range of

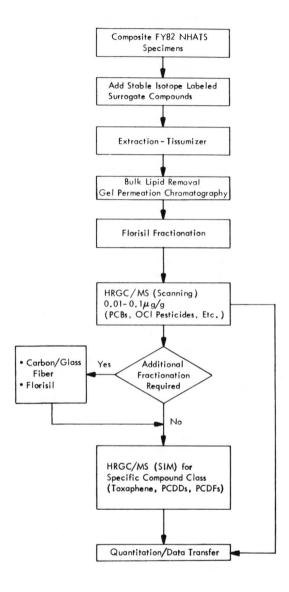

Figure 1. Flow scheme for analysis of semivolatile organic compounds in human adipose tissue.

80-550 amu was repetitively scanned every 1.7 sec. Mass spectra were acquired and stored using a Finnegan INCOS 2300 data system.

Specific analyses for toxaphene and for PCDD and PCDF required further sample extract cleanup and fractionation. Charcoal/glass fiber columns were prepared from 600 mg of Whatman CF/D glass fiber filters and 50 mg of Amoco PX-21 carbon [9]. The Florisil™ column extracts were combined and then diluted into 5 ml of cyclohexane/methylene chloride (1/1) and were transferred to the columns. The sample vials were rinsed with two 5-ml portions of the cyclohexane/methylene chloride (1/1) solvent and added to the columns. The flow rate was adjusted to 3-5 ml/min. Then 75 ml of cyclohexane/ methylene chloride (1/1) solvent was added to each column, followed by 50 ml of methylene chloride/methanol/benzene (70/25/5). Toxaphene was collected in the cyclohexane/ methylene chloride eluate. This eluate was concentrated and fractionated on deactivated Florisil™ (13 g) using a solution of 10% diethyl ether in hexane to separate toxaphene from potential interferences [10]. The flow through the columns was reversed, and 40 ml of toluene was added to each column to elute the PCDD and PCDF.

The respective fractions were analyzed for toxaphene and for PCDD and PCDF using HRGC/MS-SIM (selected ion monitoring) techniques to enhance method sensitivity for these specific compound classes. Multiple ions (including m/z 231, 233, 235, 269, 271, 273, 305, 307, 309, 327, 329, 331, 341, 343, and 345) were monitored to determine the presence of toxaphene. These ions were selected after analyzing a standard solution of toxaphene by HRGC/full scan mass spectrometry. PCDD and PCDF were detected by monitoring two ions of the characteristic molecular clusters for each of tetra through octachloro homologs and the respective surrogates.

Analysis of the extracts for the PCDD and PCDF demonstrated that the higher chlorinated compounds (hexa through octa) had not been quantitatively recovered from the Florisil™ column fractionation. Thus, an alternate cleanup procedure was necessary to achieve analytical data for the hexa- through octachloro-PCDD and PCDF. Approximately 10% (1-2 g original weight) of each sample had been reserved following the GPC step. This aliquot was taken through a carbon cleanup column consisting of 18% Carbopak™ C on Celite™ 545 [11, 12].

The sample extracts were added to the Carbopak/Celite columns with several rinses of hexane. The columns were eluted with 1 ml of cyclohexane/methylene chloride (1/1), 1 ml of methylene chloride/methanol/benzene (70/25/5), and 20 ml of toluene. The toluene fraction was concentrated and analyzed by HRGC/MS-SIM for PCDD and PCDF.

Volatile Organic Compounds

The analytical procedure for determination of volatile organic compounds in the human adipose tissue samples was based on a dynamic headspace purge and trap HRGC/MS technique. Figure 2 provides a schematic of the analytical system. The frozen composited adipose tissue samples were placed in a specially designed Wheaton™ purge chamber along with 80 ml of volatile organic free water. This mixture was spiked with 1 μ g each of several internal standards prepared in a solution of tetraglyme. This internal standard spiking solution contained 1-chloro-2-bromopropane, methylene chloride-d$_2$, chloroform-d, 1,1,2,2-tetrachloroethane-d$_2$, benzene-d$_6$, chlorobenzene-d$_5$, toluene-d$_8$, ethylbenzene-d$_{10}$, p-xylene-d$_{10}$, and 1,4-dichlorobenzene-d$_4$. The spiked aqueous mixture was allowed to equilibrate for 30 minutes before proceeding with the analysis.

The Wheaton vessel was connected to a hot water circulating bath (Haake, No. F4391) maintained at 95° C. Approximately 5.0 min was required for the solution within the vessel to reach the maximum purge temperature. The vessel was placed on a magnetic stirrer (Ace, 12064-08) and a 1.0-in. Teflon™ stirring bar was used to agitate the solution. Helium was directed into the vessel to displace the headspace at 40.0 ml/min. All metal gas carrier lines after the vessel outlet were wrapped with heat tape maintained at 150° C to prevent condensation of the target analytes and internal standards.

The effluent from the vessel line flowed into a column equipped with a stopcock and frit which contained 1.0 ml of volatile-free water. This column was used as a condenser to remove excess moisture from the purge gas. The outlet line from this purge tower was attached to the Carle valve. The Carle valve was attached to a glass-lined U tube (1/8 in. i.d.) packed with a 1.0-in. plug of Tenax-GC™ (80-100 mesh). Glass wool was used to maintain the position of the Tenax-GC in the center of the U-tube.

The U-tube was rapidly heated (approximately 5-8 sec) to 250° C. A resistance circuit with a thermocouple was used to heat and regulate the temperature of the U-tube. In the purge mode the Carle valve directed the purge gas and analytes into the U-tube, which was at ambient temperature. The analytes were trapped and the purge gas vented. The helium carrier gas was directed onto the HRGC column during the purge mode.

After the purge time had elapsed, the Carle valve was switched to the desorb mode and the U-tube was heated to flash volatize the analytes. The helium carrier gas was then routed through the U-tube in the opposite direction of the purge mode and directed onto the HRGC column.

The volatile organic compounds were analyzed using a Finnigan 9610 gas chromatograph and a Finnigan 4000 quadrupole mass spectrometer equipped with an INCOS data system. Separation of the volatile organic compounds was achieved with a Durabond DB-5 fused silica capillary column, 30 m x 0.25 mm,

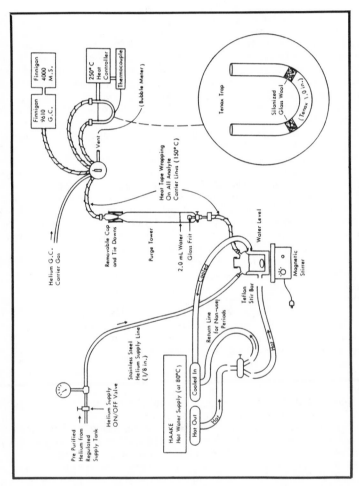

Figure 2. Schematic of the dynamic headspace purge and trap HRGC/MS analysis system.

0.25 μm film thickness (J&W Scientific, Rancho Cordova, CA).
The capillary column was routed directly into the ion source.
The helium carrier gas was adjusted at 12 psi head pressure.
The gas chromatograph was equipped with a Grob type split/
splitless injector. The effluent from the Tenex-GC adsorbent
trap was adjusted to 5-10 ml/min and directed into the Grob injector using a syringe needle attached to the stainless steel
tubing from the absorbent trap. The injector was operated in
the split mode with a 10:1 split ratio.

The gas chromatograph was held isothermal at 30° C for 5
min and then programmed at 6° C/min up to 125° C where it was
held for 10 min. Mass spectral data was acquired across the
mass range of 35-275 amu every 2-3 sec for 20 min from initiation of the program. The HRGC column was programmed to 200° C
between sample analyses to remove potential interferences for
the next analysis.

Trace Elements

The two multielement analysis techniques, inductively coupled plasma-atomic emission spectrometry (ICP-AES) and neutron
activation analysis (NAA), were evaluated for the determination
of trace elements in human adipose tissue samples. Nine adipose tissue specimens were randomly selected from the FY 82
NHATS repository. The criteria for selecting the specimens required that ample mass was available and that the tissues were
primarily fatty materials.

ICP-AES. Approximately 0.5-g aliquots of adipose tissue
were fortified with 10 μg of an internal standard, yttrium (Y),
and were then digested with 4 ml of a 50% (v/v) solution of nitric acid at an elevated temperature (110° C) for approximately
2 hours. The digested samples were diluted with deionized
water to a final weight of approximately 10 g and were analyzed
using a Jarrell-Ash Model 1155A inductively coupled argon emission spectrometer.

NAA. The neutron activation analyses were performed by
General Activation Analysis, Inc., in San Diego, California,
using 4096 and 8192 channel gamma ray spectrometer systems
equipped with Ge(Li) detectors after irradiation in a
TRIGA[TM] Mark 1 reactor. The adipose tissue samples were
weighed and sealed in polyethylene vials prior to irradiation
in the reactor. The very short-lived isotopes were determined
from approximately 1.0-g aliquots of each sample 1 min after
irradiation at a flux of 2.5 x 10^{12} n/cm²sec for 1 min.
The short-, medium-, and long-lived isotopes were determined
from approximatley 10-g aliquots of each sample 1 hr, 1 day, 1

week, and 3 weeks after irradiation for 30 min at a flux of 1.8 $\times 10^{12}$ n/cm^{2}sec.

RESULTS

This study represents a major step in the advancement of EPA's National Human Monitoring Program to monitor exposure of the general United States population to toxic organic chemicals. The data base for the number of specific xenobiotic organic compounds and trace elements detected in adipose tissue has been expanded. A summary of the results from the analyses of the FY 82 adipose tissue composites for general semivolatile organic compounds, PCDD and PCDF, volatile organic compounds, and trace elements is provided below.

Semivolatile Organic Compounds

The predominant compounds identified with the semiovolatile organic analysis procedures were noted to be the organochlorine pesticides and PCB, which have previously been monitored through PGC/ECD techniques. The HRGC/MS method, however, provides an additional confidence level for determination, since identification is based on matches of both retention time and mass spectra. In addition, the detail on PCB levels has been expanded as a result of the identification of specific degrees of chlorination (homologs) and the quantification of individual responses. Previous analyses for PCB in the NHATS monitoring program based on the PGC/ECD method have resulted in semiquantitative data based on a single response.

Quantitative data for organochlorine pesticides, polychlorinated biphenyls, chlorobenzenes, phthalate esters, phosphate triesters, and polynuclear aromatic hydrocarbons were determined for each composite sample prepared. Table 1 summarizes the incidence of detection of selected semivolatile organic compounds and the range of concentrations measured based on extractable lipid content.

The feasibility of determining other halogenated aromatic compounds, including polybrominated biphenyls, polychlorinated terphenyls, and polychlorinated diphenyl ethers using this method, was demonstrated through the analysis of spiked adipose tissue samples. However, these compounds were not detected in any of the composited FY 82 NHATS samples at concentrations as low as 0.010 to 0.050 μg/g.

The samples representing the 45+ age category were also analyzed for toxaphene by HRGC/MS-SIM. Toxaphene was qualitatively identified in 12 of the 14 samples analyzed. Quantification was not achieved, however, due to the complexity of the response, but was estimated to be less than 0.10 μg/g.

Table 1.   Incidence of Detection of Target Semivolatile Organic
           Compounds in the NHATS FY 82 Composite Specimens.

Compound	Frequency of Observation[a] (%)	Range of Observed Lipid Concentration (ng/g)
Dichlorobenzene	9	ND (9)[b] - 57
Trichlorobenzene	4	ND (9) - 21
Naphthalene	40	ND (9) - 63
Diethyl Phthalate	42	ND (10) - 970
Tributyl Phosphate	2	ND (44) - 120
Hexachlorobenzene	76	ND (12) - 1300
β-BHC	87	ND (19) - 570
Phenanthrene	13	ND (9) - 24
Di-n-butyl Phthalate	44	ND (10) - 1700
Heptachlor Epoxide	67	ND (10) - 310
trans-Nonachlor	53	ND (18) - 520
p,p'-DDE	93	ND (9) - 6800
Dieldrin	31	ND (44) - 4100
p,p'-DDT	55	ND (9) - 540
Butylbenzyl Phthalate	69	ND (9) - 1700
Triphenyl Phosphate	36	ND (18) - 850
Di-n-octyl Phthalate	31	ND (9) - 850
Mirex	13	ND (9) - 41
tris(2-chloroethyl)Phosphate	2	ND (35) - 210
Total PCB	83	ND (15) - 1700
Trichlorobiphenyl	22	ND (9) - 33
Tetrachlorobiphenyl	53	ND (9) - 93
Pentachlorobiphenyl	73	ND (21) - 270
Hexachlorobiphenyl	73	ND (19) - 450
Heptachlorobiphenyl	53	ND (19) - 390
Octachlorobiphenyl	40	ND (20) - 320
Nonachlorobiphenyl	13	ND (18) - 300
Decachlorobiphenyl	7	ND (22) - 150

[a]Sample size = 46 composites.
[b]ND = not detected. Value in parentheses is the estimated
limit of detection.

PCDD and PCDF

The results of this phase of the broad scan analysis demonstrated that the EPA NHATS program is an effective vehicle for
documenting the exposure of the general United States population to PCDD and PCDF. The analysis of the 46 composite samples prepared from the FY 82 NHATS repository establishes the

prevalence of the 2,3,7,8-substituted tetra- through octa-chloro-PCDD and PCDF congeners.

Table 2 presents the frequency of detection, mean concentration, and lipid concentration range of detection for the tetra- through octachloro-PCDD and PCDF congeners. The data in Table 2 indicate that the 2,3,7,8-TCDD was detected in 35 of the 46 composites with an average lipid-adjusted concentration of 6.2 + 3.3 pg/g. The average concentration of the other PCDD congeners ranged from 33.5 pg/g for pentachlorodibenzo-p-dioxin (detected in 91% of the composites) up to 554 pg/g for octa- chlorodibenzo-p-dioxin (detected in 100% of the composites).

Table 2. Lipid-Adjusted Concentration of PCDD and PCDF in the
NHATS FY 82 Composite Specimens.

Compound	Frequency of Detection (%)	Mean Concentration[a] (pg/g)	Range of Detection (pg/g)
2,3,7,8-TCDD	76	6.2 + 3.3	ND (1.3)[c] - 14
1,2,3,7,8-PeCDD	91	43.5 ∓ 46.5	ND (1.3) - 5000
HxCDD[b]	98	86.9 ∓ 83.8	ND (13) - 620
1,2,3,4,7,8,9-HpCDD	98	102 ∓ 93.5	ND (26) - 1300
OCDD	100	694 ∓ 355	19 - 3700
2,3,7,8-TCDF	26	15.6 ∓ 16.5	ND (1.3) - 660
2,3,4,7,8-PeCDF	89	36.1 ∓ 20.4	ND (1.3) - 90
HxCDF[b]	72	23.5 ∓ 11.6	ND (3.0) - 60
1,2,3,4,6,7,8-HpCDF	93	20.9 ∓ 15.0	ND (3.5) - 79
OCDF	39	73.4 ∓ 134	ND (1.2) - 890

[a]Mean concentration calculated using trace and positive quantifiable values.
[b]Reference compounds not available to specify isomers.
[c]ND = not detected. Value in parentheses is the estimated limit of detection.

The data demonstrate some differences in PCDD levels for the three age groups evaluated (Figure 3). The PCDF were generally detected less frequently and were present at lower concentration than the PCDD. Obvious trends in the levels of the PCDF congeners with respect to age were not observed. The mean values for the PCDD and PCDF data are comparable to values that have been reported for other studies on adipose tissue samples, from the United States [13], Sweden [14], and Canada [15,16].

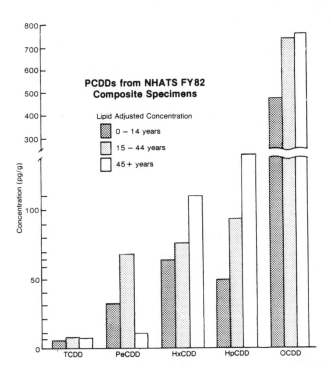

Figure 3. PCDD distribution in the general United States population as a function of age group.

Volatile Organic Compounds

The exposure of the general United States population to
volatile organic compounds has not been previously addressed
through a national sampling of biological matrices (breath,
blood, or tissue). Studies have been conducted, however, to
determine the effects of exposure to specific chemical solvents, monomers such as vinyl chloride and styrene in the plastics industry, and anesthetics [17-20]. The fact that blood
and breath levels of volatile organics can be detected at de-
clining levels from several hours to several days after a specific exposure incident indicates tissue retention [17,21,
22]. Human adipose tissue has been evaluated as a depot for
storage and release of volatiles in specific exposure studies
of workers to styrene and ethylbenzene in the polymerization
industry [17,18,19].

The broad scan analysis with the FY 82 NHATS specimens demonstrates that adipose tissue may be useful in assessing human exposure to volatile as well as semivolatile compounds. Quantitative data for 17 halogenated and/or aromatic volatile compounds were determined from the analyses of each of the 46 adipose composites. Predominant volatile organic compounds in the composited human adipose tissues quantified in this study included chloroform, 1,1,1-trichloroethane, benzene, tetrachloroethane, toluene, chlorobenzene, ethylbenzene, styrene, 1,1,2,2-tetrachloroethane, 1,4-dichlorobenzene, xylenes, and ethylphenol. Several compounds including styrene, the xylene isomers, 1,4-dichlorobenzene, and ethylphenol were detected in all the composite specimens. Table 3 presents the incidence of detection of selected target analytes and the range of concentration observed.

Table 3. Incidence of Detection of Target Volatile Organic Compounds in the NHATS FY 82 Composite Specimens.

Compound	Frequency of Observation (%)	Wet Tissue Concentration (ng/g)
Chloroform	76	ND (2)[a] - 580
1,1,1-Trichloroethane	48	ND (17) - 830
Bromodichloromethane	0	ND (21)
Benzene	96	ND (4) - 97
Tetrachloroethane	61	ND (3) - 94
Dibromochloromethane	0	ND (1)
1,1,2-Trichloroethane	0	ND (1)
Toluene	91	ND (1) - 250
Chlorobenzene	96	ND (1) - 9
Ethylbenzene	96	ND (2) - 280
Bromoform	0	ND (1)
Styrene	100	8 - 350
1,1,2,2-Tetrachloroethane	9	ND (1) - 8
1,2-Dichlorobenzene	63	ND (0.1) - 2
1,4-Dichlorobenzene	100	12 - 500
Xylene	100	18 - 1400
Ethylphenol	100	0.4 - 400

[a]ND = not detected. Value in parentheses is the estimated limit of detection.

Trace Elements

Analysis of selected adipose tissue specimens using the two multielement techniques was limited to nine specimens. This phase of the study was intended to evaluate the multielement analysis technique and to determine whether toxic trace element data in adipose tissue might be of interest to EPA. A total of 18 elements were detected and quantified using ICP-AES and NAA techniques.

Elements detected by ICP-AES included aluminum, boron, calcium, iron, magnesium, sodium, phosphorus, tin, and zinc. Elements detected by NAA included bromine, chlorine, cobalt, iron, iodine, potassium, sodium, rubidium, selenium, silver, and zinc.

Very little information is available in the open literature regarding the levels of specific elements in human adipose tissue. The most significant source of information presented on human adipose tissue levels was found in a report prepared for the International Commission on Radiological Protection (ICRP) [23]. The ICRP report summarizes elemental composition based on total body organ and tissue type for what is referred to as "reference man." The information presented in that report was taken from several literature sources, but much of it is based on activities completed at the Oak Ridge National Laboratory and University of Tennessee from the mid 1950s to the mid 1960s. The report does not specify the exact analytical procedures used to obtain the data, although some general references are made with respect to colorimetric, atomic emission, atomic absorption, and DC-arc plasma emission techniques.

Table 4 presents a comparison of the range of concentrations observed for specific elements from the NHATS specimens in this study and the estimates presented for "reference man" in the ICRP report. The ICRP report specifies that "reference man" consists of a total mass of 70 kg ($\sim$ 150 lb), with as much as 21% of 15 kg of total body mass consisting of adipose tissue. The general term "adipose tissue" in the ICRP report includes subcutaneous adipose, adipose surrounding specific organs such as the kidneys or intestines, and interstitial adipose interspersed among the cells of an organ and yellow marrow. In general, the data generated by the two multielement techniques are close to the data for "reference man." The most obvious differences in values for "reference man" and the NHATS specimens are noted for boron, silver, and tin.

SUMMARY

The broad scan analysis has resulted in the development and preliminary evaluation of HRGC/MS methods for the measurement of semivolatile and volatile organic compounds at concentrations ranging from 0.001 to 2 μg/g in human adipose tissues. Specific procedures based on SIM techniques have provided qualitative analysis for complex analytes such as toxaphene. The

Table 4. Comparison of Elements Detected in the NHATS FY 82 and the ICRP Reference Man.

Element	Reported Concentration (μg/g)	
	NHATS FY 82 Specimens	ICRP Reference Man[a]
Aluminum (Al)	ND (0.63) - 4.3	0.35
Boron (B)	ND (0.32) - 22	0.073
Bromine (Br)	0.33 - 2.4[b]	0.43
Calcium (Ca)	15 - 98	23
Chlorine (Cl)	360 - 1500[b]	1200
Cobalt (Co)	0.034 - 0.079[b]	0.024
Gold (Au)	ND - 0.0030[b]	<0.33
Iodine (I)	ND (1.4) - 13[b]	c
Iron (Fe)	3.0 - 36	2.4
	3.5 - 26[b]	
Magnesium (Mg)	6.5 - 25	20
Phosphorus (P)	130 - 220	160
Potassium (K)	52 - 270[b]	320
Rubidium (Rb)	ND - 0.27[b]	c
Selenium (Se)	ND - 0.56[b]	c
Silver (Ag)	ND - 0.38	0.0013
Sodium (Na)	150 - 540	510
	240 - 1200[b]	
Tin (Sn)	4.6 - 15	0.047
Zinc (Zn)	1.1 - 6.0	1.8
	1.4 - 4.5[b]	

[a]Snyder, W.S., M.J. Cook, E.S. Nasset, L.R. Karhausen, G.P. Howells, and I.H. Tipton. "Report of the Task Group on Reference Man," International Commission on Radiological Protection No. 23, (Pergamon Press, 1975). pp. 273-334.
[b]Values from NAA. All other values for the NHATS specimens were observed by ICP-AES.
[c]No estimate provided.

sensitivity of the SIM technique has also been applied to the determination of parts per trillion (picogram/gram) quantities of specific PCDD and PCDF congeners. Two multielement analysis techniques were evaluated in an effort to determine the levels of toxic trace elements in adipose tissue.

The analysis of the composited FY 82 NHATS specimens has resulted in the identification of additional specific xenobiotic compounds beyond chlorinated pesticides and PCB. The prevalence of specific volatile organic compounds and PCDD and PCDF in the adipose tissue of the general United States population was demonstrated.

The quantitative data for each of the specific compounds are currently being evaluated by statistical analysis techniques to determine if significant trends exist as a result of geographic location, age group, sex, or race.

The analytical methods will require further modification and validation before establishing procedures for routine monitoring of the specific compounds through the NHATS program. EPA/OTS has initiated a study to address the comparability of the data for organochlorine pesticides and PCB generated by the HRC/MS method and the PGC/ECD method used prior to the FY 82 collection.

Although as many as 50-60 volatile and semivolatile organic compounds were identified in the composited adipose tissue samples, the HRGC/MS data contain a significant amount of mass spectral information for which compound identifications have not been assigned. In order to obtain maximum information from these broad scan analyses, a program has been established through EPA/OTS to address these unidentified responses.

ACKNOWLEDGMENT

The research described in this chapter was funded by the U.S. Environmental Protection Agency under contracts 68-02-3938 to Midwest Research Institute and 68-02-4243 to Battelle Columbus Laboratories.

REFERENCES

1. Yobs, A. R. "The National Human Monitoring Program for Pesticides," Pestic. Monitor. J. 5:46 (1971).

2. Sherma, J., and M. Beroza. "Analysis of Pesticide Residues in Human and Environmental Samples: A Compilation of Methods Selected for Use in Pesticide Monitoring Programs," EPA Report-600/8-80-038, Section 5,A,(1),(a), pp. 1-19 (1980).

3. Kutz, F. W., S. C. Strassman, and J. F. Sperling. "Survey of Selected Organochlorine Pesticides in the General Population of the United States: Fiscal Year 1970-1975," Annals N.Y. Acad. Sci. 31:60-68 (1979).

4. Kutz, F. W., A. R. Yobs, S. C. Strassman,and J. R. Viar, Jr. "Pesticides in People, Effects of Reducing DDT Usage on Total DDT Storage in Humans," Pestic. Monitor. J. 11(2):61-63 (1977).

5. Kutz, F. W., A. R. Yobs, and S. C. Strassman. "Organo-chlorine Pesticide Residues in Human Adipose Tissue," Bull. Sco. Pharmacol. Environ. Pathol. 4:17-19 (1976).

6. Kutz, F. W., G. W. Sovocool, S. Strassman,and R. G. Lewis. "trans-Nonachlor Residues in Human Adipose Tissue," Bull. Environ. Cont. Toxicol. 16:9-14 (1976).

7. Lewis, R. G., and G. W. Sovocool. "Identification of Polybrominated Biphenyls in the Adipose Tissues of the General Population of the United States," J. Anal. Toxicol. 6:196-198 (1982).

8. Wright, L. H., R. G. Lewis, H. L. Crist, G. W. Sovocool, and J. M. Simpson. "The Identification of Polychlorinated Terphenyls at Trace Levels in Human Adipose Tissue by Gas Chromatography/Mass Spectrometry," J. Anal. Toxicol. 2:76-79 (1978).

9. Smith, L. M., D. L. Stalling, and J. L. Johnson. "Determination of Part-per-Trillion Levels of Polychlorinated Dibenzofurans and Dioxins in Environmental Samples," Anal. Chem. 56:1830-1842 (1984).

10. Zell, M., and K. Ballschmiter. "Baseline Studies of the Global Pollution II. Global Occurrence of Hexachloroben-zene (HCB) and Polychlorocamphenes (Toxaphene) (PCC) in Biological Samples," Fresenius Z. Anal. Chem. 300:387-402 (1980).

11. Kleopfer, R. D., K. T. Yue, and W. W. Bunn. "Determination of 2,3,7,8-Tetrachlorodibenzo-p-Dioxin in Soil," in Chlorinated Dioxins and Dibenzofurans in the Total Environment II, L. H. Keith, C. Rappe, and G. Choudhary, Eds. (Stoneham, MA: Butterworth Publishers, 1985), pp. 367-376.

12. Smith, R. M., P. W. O'Keefe, K. M. Aldous, D. R. Hilker, and J. E. O'Brien. "2,3,7,8-Tetrachlorodibenzo-p-Dioxin in Sediment Samples from Love Canal Storm Sewers and Creeks," Environ. Sci. Tech. 17: 6-10 (1983).

13. Schecter, A., and J. J. Ryan. "Dioxin and Furan Levels in Human Adipose Tissue from Exposed and Control Populations," 189th National ACS Meeting Symposium on Chlorinated Dioxins and Dibenzofurans in the Total Environment III, Miami, Florida 4/29-5/3 1985. Division of Environmental Chemistry, ACS 25:160-163, Paper No. 56 (1985).

14. Nygren, M., M. Hansson, C. Rappe, L. Domellof, and L. Hardell. "Analysis of Polychlorinated Dibenzo-p-Dioxins and Dibenzofurans in Adipose Tissue from Soft-Tissue Sarcoma Patients and Controls," 189th National ACS Meeting Symposium on Chlorinated Dioxins and Dibenzofurans in the Total Environment III, Miami, Florida 4/29-5/3 1985. Division of Environmental Chemistry, ACS 25:160-163, Paper No. 55 (1985).

15. Ryan, J. J., D. T. Williams, B. P. Y. Lau, and T. Sakuma. "Analysis of Human Fat Tissue for 2,3,7,8- Tetrachlorodibenzo-p-Dioxin and Chlorinated Dibenzofuran Residues," in Dioxins and Dibenzofurans in the Total Environment II, L. H. Keith, C. Rappe, and G. Choudhary, Eds. (Stoneham, MA: Butterworth Publishers, 1985), pp. 205-214.

16. Ryan, J. J., A. Schecter, R. Lizotte, W. -F. Sun, and L. Miller. "Tissue Distribution of Dioxins and Furans in Humans from the General Population," Chemosphere 14:929-932 (1985).

17. Wolff, M. S. "Evidence of Existence in Human Tissues in Monomers from Plastic and Rubber Manufacture," Environ. Health Perspect. 17:183-187 (1976).

18. Engstrom, K. "Styrene," in Biological Monitoring and Surveillance of Workers Exposed to Chemicals, A. Antero, V. Riihimäki, and H. Vainio, Eds. (Washington, D.C.: Hemisphere Publishing Corporation, 1984), pp. 99-110.

19. Engstrom, J., and V. Riihimäki. "Distribution of m-Xylene to Subcutaneous Adipose Tissue in Short Term Experimental Human Exposure," Scand. J. Work Environ. Health 5:126-134 (1979).

20. Wolff, M. S., S. M. Daum, W. V. Loumer, I. J. Selikoff, and B. B. Aubrey. "Styrene and Related Hydrocarbons in Subcutaneous Fat from Polymerization Workers," Toxicol. Environ. Health 2:997-1005 (1977).

21. Corbett, T. H. "Retention of Anesthetic Agents Following Occupational Exposure," Anesth. Anal. 52:614 (1973).

22. Whitcher, C. E., E. N. Cohen, and J. R. Trudell. "Chronic Exposure to Anesthetic Gases in the Operating Room, Anesth. 35:349 (1971).

23. Snyder, W. S., M. J. Cook, E. S. Nasset, L. R. Karhausen, G. P. Howells, and I. H. Tipton. "Report of the Task Group on Reference Man," International Commission on Radiological Protection No. 23 (Pergamon Press, 1975).

CHAPTER 15

RESULTS FROM THE FIRST THREE SEASONS OF THE TEAM STUDY:
PERSONAL EXPOSURES, INDOOR-OUTDOOR RELATIONSHIPS, AND
BREATH LEVELS OF TOXIC AIR POLLUTANTS MEASURED FOR 355
PERSONS IN NEW JERSEY

Lance A. Wallace, Edo D. Pellizzari, Ty D. Hartwell, Charles M.
Sparacino, Linda S. Sheldon, and Harvey Zelon

INTRODUCTION

EPA's TEAM (Total Exposure Assessment Methodology) Study
was designed to develop and demonstrate methods to measure
human exposure to toxic substances in air, food, and drinking
water, and to measure biological fluids (breath, blood, urine)
for the same compounds to determine body burden. A first phase
to field-test the methods was completed in 1981 [1,2,3]. Methods developed or demonstrated in Phase I included a personal
monitor employing Tenax^{TM} cartridges, a spirometer for collecting expired air on Tenax cartridges, and a statistical design with field-tested questionnaires for the present study.
The objective of the second phase [4,5,6] was to estimate
the distribution of exposures to target substances for the entire population of an industrial/chemical manufacturing area.
A total of 20 toxic, carcinogenic, or mutagenic organic compounds was measured in the air and drinking water of 355 residents of Bayonne and Elizabeth, New Jersey, between September 3
and November 23, 1981. The participants were selected from
over 10,000 residents screened by a probability sampling technique to represent 128,000 persons (over the age of 7) who live
in the two neighboring cities, which include extensive chemical
manufacturing and petrochemical refining activities.
One hundred geographic areas throughout the two cities were
selected for monitoring. Each participant carried a personal
sampler with him during his normal daily activities for 2 con-
secutive 12-hour periods. (One resident in each of the 100

areas had an identical sampler operating in the back yard for the same two 12-hour periods.) All participants also collected two drinking water samples. At the end of the 24-hour sampling period, all participants gave a sample of exhaled breath, which was analyzed for the same compounds. All participants also completed a questionnaire on their occupations and activities during the sampling period. An extensive quality assurance program was carried out on all sampling/analysis activities.

Return visits were made in the summer of 1982 to 160 persons, and in February 1983 to 50 members of the original group. Similar procedures were followed on all three visits.

MEASUREMENT METHODS

Air

Personal and outdoor air samples were collected on Tenax cartridges for 12-hour periods. A Dupont^{TM} pump pulled air at 30 ml/min ($\sim$22 1 sampling volume) across the 1.5 cm i.d. cartridge, which contained 6 cm ($\sim$2 g) of 40/60 mesh purified Tenax. Cartridges were analyzed by thermal desorption and cryofocusing of the organic vapors [7], followed by capillary gas chromatography/mass spectrometry/computer analysis (GC/MS/COMP).

Breath

Breath samples were collected by a specially designed spirometer consisting of a humidified supply of pure air, a 40-1 Tedlar^{TM} bag to collect expired air, and two Nutech^{TM} pumps to pull the expired air across two Tenax cartridges. After the first bag is partially filled with pure air, the subject uses the two-way mouthpiece to inhale from the bag and exhale into the second bag. The pumps pull the expired air across the 2 Tenax cartridges, which are then stored at -20°C. Analysis is by GC/MS/COMP. Background contamination of the bags is reduced to acceptable levels by flushing with helium at least 10 times over a period of days before use.

Water

After a 20-second run, drinking water samples were collected in the morning and evening from the kitchen tap in 40-ml

Teflon™-capped amber glass vials containing 5 mg sodium thiosulfate. For analysis, a purge-and-trap technique was used (modified from [8]). Purgable organics were swept onto a Tenax cartridge from a specially designed all-glass 25-ml purge device connected to a short gas chromatographic column used to trap compounds of interest. Aromatics were then analyzed by flame ionization and halocarbons by a Hall electrolytic conductivity detector.

Quality Assurance

Blank samples and control samples spiked with all 20 target compounds were kept at the laboratory and shipped to the field to determine background contamination levels, recovery efficiencies, and effects of transportation and storage. Duplicate air, water, and breath samples were collected and analyzed at the primary laboratory (Research Triangle Institute) and QA laboratory (IIT Research Institute) to determine intralaboratory and interlaboratory precision. Deuterated benzene was loaded on all duplicate cartridges to determine unambiguously the extent of losses during sampling operations. Periodic audits were carried out by EPA's Environmental Monitoring Systems Laboratory at Research Triangle Park (EMSL-RTP).

RESULTS

About 4400 of the 5200 target households were contacted and information was obtained on 11,414 household residents. These data were employed in the second stage to select a sample of participants. The sample was weighted to overrepresent certain high potential exposure groups. About 58% of the eligible residents (all persons 7 or older not living in group quarters) in each city agreed to participate fully in the study (Table 1). Limited follow-up studies on nonrespondents showed no outstanding differences from respondents.

About 1,950 air, breath, and water samples were collected and chemically analyzed during the fall-of-1981 visit, 800 during the summer of 1982, and 250 during the winter of 1983. An additional 980 quality control samples (duplicates, spikes, and blanks) were analyzed during the fall of 1981, 400 during the summer of 1982, and 120 during the winter of 1983.

Table 1.   Results of Two-Stage Probability Sampling: Team
Study, All Three Seasons.

	Bayonne	Elizabeth
Stage I: Screening		
Households screened	2063	3145
Households completing questionnaire	1788 (87%)	2638 (84%)
Persons providing data	4687	6727
Stage II: Monitoring		
Eligible persons	266	345
Persons completing data collection		
Fall 1981	154	201
Summer 1982	70	87
Winter 1983	22	27

Quality Assurance Results

First season (fall) results from 155 blank cartridges showed low backgrounds (corresponding to < 2 $\mu g/m^3$) except for
benzene (5 + 3 $\mu g/m^3$).   (Mean backgrounds for each batch of
Tenax cartridges were subtracted from the measured amounts on
field cartridges from that batch.)   The results from 201 spiked
control cartridges showed recovery efficiencies ranging from 85
to 110%.   The deuterated benzene results showed consistently
acceptable losses of 5-15%.
Second season (summer) quality assurance results indicated
widespread contamination of the Tenax cartridges.   This was
traced to renovations in the New Jersey hotel where the cartridges were kept during the summer sampling trip.   Third
season (winter) quality assurance results indicated unusually
clean Tenax batches.
First season results from 134 pairs of duplicate personal air
samples and 34 duplicate outdoor air samples analyzed at the
primary laboratory showed median coefficients of variation
(C.V.) ranging from 24 to 17%, except for benzene (36-47%).
Thirty duplicate breath samples had median C.V.'s of 16-46%.
Quality assurance samples analyzed at different laboratories
had larger median C.V.'s of 30-40% (90 air samples) and 30-50%
(49 breath samples).

Percent Detected

For each of the 19-20 target chemicals, the estimated percent of samples above quantifiable concentrations for all three
seasons is shown for breath and air samples (Table 2) and for
water samples (Table 3).

Table 2. Target Compounds Sorted by Percent Measurable in
Breath and Air Samples: All Three Seasons.

Compounds	Range of % Measurable[a]
Ubiquitous	
Benzene	55 - 100
Tetrachloroethylene	66 - 100
Ethylbenzene	62 - 100
o-Xylene	58 - 100
m,p-Xylene	68 - 100
m,p-Dichlorobenzene	44 - 100
1,1,1-Trichloroethane	33 - 99
Often Present	
Chloroform	4 - 92
Trichloroethylene	33 - 79
Sytrene	46 - 91
Occasionally Found	
Vinylidene Chloride	0 - 95
1,2-Dichloroethane	0 - 22
Carbon Tetrachloride	0 - 53
Chlorobenzene	2 - 40
o-Dichlorobenzene	1 - 34
Bromodichloromethane	0 - 24
Dibromochloromethane	0 - 1
Bromoform	0 - 1
Dibromochloropropane	0 - 1

[a]Percent of samples exceeding the quantifiable limit (QL) in
personal air, outdoor air, and breath samples.

Table 3.   Target Compounds Sorted by Percent Measurable in
           Water Samples: All Three Seasons.

Compounds	Range of % Measurable[a]
Ubiquitous	
Chloroform	99 - 100
Bromodichloromethane	99 - 100
Dibromochloromethane	93 - 100
Often Present	
1,1,1-Trichloroethane	46 - 50
Trichloroethylene	44 - 51
Tetrachloroethylene	43 - 53
Occasionally Found	
Vinylidene Chloride	26 - 43
1,2-Dichloroethane	1
Benzene	1 - 25
Carbon Tetrachloride	6 - 18
Bromoform	2 - 6
Chlorobenzene	0 - 1
m,p-Dichlorobenzene	0 - 3
Never Found	
Styrene	0
Ethylbenzene	0
m,p-Xylene	0

[a]Percent of samples exceeding the quantifiable limit (QL) in
personal air, outdoor air, and breath samples.

Observed Concentrations

Estimated arithmetic means and maxima of personal and outdoor air concentrations and breath levels of the most prevalent
chemicals are shown in Tables 4 and 5. Drinking water concentrations are displayed in Table 6. Distributions were right-
skewed and often close to being lognormal, with geometric standard deviations between 2.5 and 3.5 in many cases.

Since the overnight personal air samples are taken in the
subjects' homes, they may be considered essentially indoor samples. Thus, indoor-outdoor ratios can be calculated for the
matched indoor-outdoor pairs over all three seasons. These
ratios are almost always greater than one, indicating indoor
sources for all prevalent chemicals. At median concentrations,
indoor-outdoor ratios range from 1.5 to 4.0, but at the maximum
concentrations many chemicals display indoor-outdoor ratios of

Table 4. Arithmetic Means ($\mu g/m^3$) for Air and Breath Concentrations of Organic Compounds in New Jersey.

Chemical	Fall 1981 (128,000)[a] Overnight Air			Summer 1982 (109,000) Overnight Air			Winter 1983 (94,000) Overnight Air		
	Personal	Outdoor	Breath	Personal	Outdoor	Breath	Personal	Outdoor	Breath
Chloroform	8.7[b]	1.2	3.1	4.6	12.0	6.3	4.0	0.1	0.3
1,1,1-Trichloroethane	110.0	5.4	15.0	21.0	10.0	15.0	31.0	1.4	4.0
Benzene	30.0	8.6	19.0	nc[c]	nc[c]	nc[c]	nc[c]	nc[c]	nc[c]
Carbon Tetrachloride	14.0	1.2	1.3	1.2	1.0	0.4	nd[d]	nd[d]	nd[d]
Trichloroethylene	7.3	2.1	1.8	4.8	7.8	5.9	3.0	0.2	0.6
Tetrachloroethylene	11.0	3.7	13.0	9.0	4.0	10.0	13.0	1.9	11.0
Styrene	2.7	0.9	1.2	2.0	0.6	1.6	2.2	0.6	0.7
m,p-Dichlorobenzene	56.0	1.5	8.1	49.0	1.4	6.3	54.0	1.2	6.2
Ethylbenzene	13.0	3.8	4.6	7.8	3.5	4.7	11.0	3.4	2.1
o-Xylene	16.0	4.0	3.4	8.0	4.3	5.4	9.8	3.1	1.6
m,p-Xylene	55.0	11.0	9.0	19.0	11.0	10.0	29.0	8.5	4.7

[a] Population of Elizabeth and Bayonne for which estimates apply.
[b] Arithmetic means of all samples; samples below the limit of detection (LOD) assigned one-half the LOD.
[c] Not calculated - cartridges contaminated.
[d] Not detected in most samples.

Table 5.   Maximum Concentrations ($\mu g/m^3$) of Organic Compounds in Air and Breath of 350 NJ Residents.

Chemical	Personal Air[a]		Outdoor Air[b]		Breath[c]
	Night	Day	Night	Day	
Chloroform	210	140	130	230	29
1,1,1-Trichloroethane	8,300	330,000	51	470	520
Benzene	510	270	91	44	200
Carbon Tetrachloride	1,100	900	14	7.1	250
Trichloroethylene	350	1,400	61	100	30
Tetrachloroethylene	250	12,000	27	95	280
Styrene	76	6,500	11	6.3	31
m,p-Dichlorobenzene	1,600	2,600	13	57	160
Ethylbenzene	380	1,500	28	39	290
o-Xylene	750	1,800	31	19	220
m,p-Xylene	3,100	10,000	70	47	350

[a]No. of samples ~ 540 during 3 seasons.
[b]No. of samples ~ 150 during 3 seasons.
[c]No. of samples ~ 500 during 3 seasons.

Table 6.   Arithmetic Means and Maxima ($\mu g/m^3$) of Organic Compounds in New Jersey Drinking Water.

Chemical	Fall 1981 (128,000)[a]		Summer 1982 (109,000)[a]		Winter 1983 (94,000)[a]	
	Mean[b]	Max	Mean	Max	Mean	Max
Chloroform	70	170	61	130	17	33
Bromodichloromethane	14	23	14	54	5.4	16
Dibromochloromethane	2.4	8.4	2.1	7.2	1.4	3.0
1,1,1-Trichloroethane	0.6	5.3	0.2	2.6	0.2	1.6
Trichloroethylene	0.6	4.2	0.4	8.3	0.4	3.4
Tetrachloroethylene	0.4	3.3	0.4	9.3	0.4	5.0
Toluene	0.4	2.7	--	--	--	--
Vinylidene Chloride	0.2	2.4	0.1	2.5	0.2	0.9
Benzene	--	--	0.7	4.8	--	--

[a]Population of Bayonne and Elizabeth to which estimates apply.
[b]Arithmetic mean of all samples; values below the limit of detection (LOD) assigned a value one-half the LOD.

10 or 20, indicating very strong indoor sources. The ratios increase from summer to fall to winter (an example is shown for m+p-dichlorobenzene in Figure 1).

Uncertainty of Estimates

The uncertainty in the estimates of personal exposures of the target population consists of two parts: survey sampling uncertainty and measurement errors. For an unstratified sample size of 350 persons, assuming a lognormal distribution, standard sampling theory states that the estimate of the median will be 95% certain to lie between the 44th and 56th percentiles [9, p. 445]. Since our sample is stratified, the stratification design effect will be to broaden these ranges of uncertainty by a small amount. The corresponding range for the summer group of 160 persons is 41-59%; and for the winter group of 50 persons, 35-65%.

The second source of uncertainty is measurement error. Analysis of the duplicate measurements obtained during all three seasons following the method of Evans [10] has resulted in improved estimates of the frequency distribution of exposures, showing that the observed geometric standard deviations should be reduced by 5 to 20%. The correction factors by which the observed fall 1981 90th (or 75th) percentile values should be multiplied to give the estimated "true" 90th (or 75th) percentile values are listed in Table 7.

Correlations in Air

Spearman correlations were calculated for all possible pairs of the target chemicals within the overnight and daytime personal air and outdoor air samples. Correlations were high between certain groups of associated chemicals. For example, the xylene isomers and ethylbenzene, found in gasoline and paints in about the same relative proportions, had correlation coefficients exceeding 0.9 in all cases. On the other hand, chloroform and paradichlorobenzene showed little correlation with any of the other chemicals or with each other.

Correlations Between Breath and Air

Correlations of breath levels with preceding 12-hour average personal air exposures were almost always significant (p < .05) except for chloroform, for which the main route of

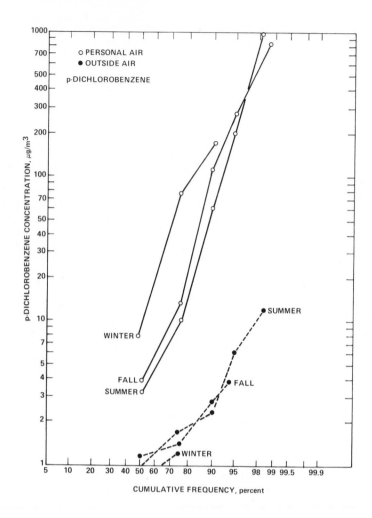

CUMULATIVE FREQUENCY, percent

Figure 1. Frequency distributions for overnight personal air
exposures and overnight outdoor air concentrations of
para-dichlorobenzene in Bayonne and Elizabeth, NJ
during three seasons.   Although outdoor concentrations show little change, indoor concentrations increase by factors of 2 to 6 between summer and winter, due perhaps to reduced air exchange in winter.
Distributions are weighted to represent the entire
target populations during the three seasons: Fall,
1981, 128,000 (N_p = Number of personal air samples
= 347; N_o = Number of outdoor air samples = 84);
Summer, 1982, 109,000 (N_p = 147; N_o = 72); Winter, 1983, 94,000 (N_p = 47; N_o = 7).

Table 7. Correction Factors Due to Measurement Errors:
          Fall 1981.

Chemical	Breath[a]	Personal Air[a]		Outdoor Air[b]	
		Night[a]	Day[a]	Night[b]	Day[b]
Chloroform	0.70	0.96	0.92	--[c]	0.87
1,1,1-Trichloroethane	0.60	0.93	0.81	0.82	0.91
Benzene	--	0.75	0.62	--	0.66
Carbon Tetrachloride	0.97	0.92	0.63	0.95	--
Trichloroethylene	0.84	0.96	0.84	0.98	0.86
Tetrachloroethylene	0.85	0.93	0.96	0.92	0.97
Styrene	0.90	0.89	0.68	--	0.77
m,p-Dichlorobenzene	0.96	0.96	0.92	0.98	0.97
Ethylbenzene	--	0.89	0.92	0.98	0.92
o-Xylene	0.55	0.74	0.92	0.95	0.92
m,p-Xylene	0.50	0.81	0.84	0.93	0.75

[a]"True" 90th percentile value/observed 90th percentile.
[b]"True" 75th percentile value/observed 75th percentile.
[c]"True" value cannot be calculated - measurement errors too
   large.

exposure is drinking water (Table 8). The chemical most highly
correlated with previous exposures was paradichlorobenzene.
     Benzene concentrations in air and breath were significantly
different for smokers and non-smokers (Figure 2).   Over the
three seasons, benzene levels in the breath of smokers averaged
6-fold increases compared to non-smokers. Styrene, xylenes,
and ethylbenzene were also elevated in the breath of smokers.
Homes with smokers had elevated levels of benzene compared to
homes without smokers, suggesting the presence of benzene in
sidestream tobacco smoke.

Effects of Residence and Activities on Exposure

     Few differences between Bayonne and Elizabeth were noted
for personal or outdoor air samples, breath samples, or drinking water samples. Persons living within 1.5 km of major point
sources showed no differences in personal air, outdoor air, or
drinking water exposures when compared to persons living far-
ther away.
     The data collected on the 24-hour activity recall question-
naires proved useful in identifying the probable sources of
high exposures. For example, participants visiting a gasoline

Table 8.   Spearman Correlations Between Breath Values and
           Preceding 12-Hour Personal Air Concentrations: TEAM
           Study.

Chemical	Fall 1981 (N $\sim$ 300)	Summer 1982 (N $\sim$ 130)	Winter 1983 (N = 47)
m,p-Dichlorobenzene	.54*	.38*	.61*
Tetrachloroethylene	.46*	.23*	.37*
Trichloroethylene	.38*	--a	.35*
m,p-Xylene	.32*	.27*	.48*
Ethylbenzene	.33*	.22*	.44*
1,1,-Trichloroethane	.28*	.28*	.32*
o-Xylene	.26*	.22*	.45*
Benzene	.21*	ncb	nc
Styrene	.19*	.20*	--a
Chloroform	--a	--a	--a

*Significantly different from 0 (p<.5).
aCorrelation less than 0.2.
bNot calculated - did not meet quality assurance standards.

service station on the day they were monitored showed significantly higher levels of benzene (but not other chemicals) in
personal air and breath samples.   Similarly, those visiting a
dry cleaner showed significantly higher levels of tetrachloroethylene in air and breath.   Smokers showed significantly higher levels of benzene, xylene, styrene, and ethylbenzene in air
and breath.   Persons exposed to paint, plastics, and chemical
plants had higher levels of ethylbenzene, styrene, and xylene
isomers in air and breath.   Other potential sources as identified by stepwise regressions of the log-transformed data are
listed in Table 9.

DISCUSSION

Personal and Indoor Air

     Personal air exposures to   all 11 of the most prevalent
chemicals were greater - often much greater - than would have
been predicted from outdoor monitoring alone.   The major cause
of these higher exposures appears to be in the home, since
overnight concentrations in the home were consistently greater
than in the adjoining backyard.   Indoor-outdoor ratios increased from summer to fall to winter, a finding consistent

Table 9. Activities, Occupations, or Household Characteristics
Associated with Significantly Increased Exposures in
Air or Breath to Eleven Prevalent Chemicals in New
Jersey.

Chemical	Rank	Activity[a]	p[b]
Benzene	1	Smoking	0.00001
	2	Having a smoker in the home	0.0006
	3	Being exposed to smokers	0.02
	4	Visiting a dry cleaners	0.03
	5	Traveling in a car	0.04
Styrene	1	Smoking	0.00001
	2	Having a smoker in the home	0.0001
	3	Working at a plastics plant	0.0005
	4	Exposed to paints	0.002
	5	Working at/being in a paint store	0.005
	6	Working at a chemical plant	0.007
	7	Building scale models	0.007
	8	Painting as a hobby	0.009
	9	Being nonwhite	0.01
	10	Metalworking	0.02
	11	Working with degreasers	0.03
Ethylbenzene	1	Smoking	0.0001
	2	Exposed to high dust/particle levels	0.0002
	3	Having a smoker in the home	0.0006
	4	Working with solvents	0.001
	5	Wood processing	0.001
	6	Working at a service station	0.002
	7	Having a chemical worker in the home	0.002
	8	Employed	0.002
	9	Living in a home less than 1 year	0.003
	10	Pumping gas	0.005
	11	Metalworking	0.005
	12	Working at a scientific lab	0.005
	13	Refinishing furniture as a hobby	0.01
	14	Working with dyes	0.01
	15	Having a metal worker in the home	0.02
	16	Working with odorous chemicals	0.02
	17	Having a furniture refinishing hobbyist in the home	0.04

Table 9.   (Continued).

Chemical	Rank	Activity[a]	p[b]
m,p-Xylene	1	Employed	0.0001
	2	Smoking	0.0001
	3	Wood processing	0.0001
	4	Working at a service station	0.0001
	5	Having a chemical worker in the home	0.0001
	6	Working with solvents	0.0003
	7	Having a smoker in the home	0.0006
	8	Living in an old home (more than 10 years)	0.002
	9	Living in a home less than 1 year	0.003
	10	Pumping gas	0.006
	11	Metalworking	0.008
	12	Exposed to high dust/particle levels	0.01
	13	Having a furniture refinishing hobbyist in the home	0.02
	14	Furniture refinishing	0.03
o-Xylene	1	Wood processing	0.0001
	2	Employed	0.0001
	3	Working with solvents	0.0008
	4	Working with odorous chemicals	0.001
	5	Pumping gas	0.002
	6	Metalworking	0.003
	7	Having a chemical worker in the home	0.003
	8	Having a smoker in the home	0.005
	9	High dust/particle levels	0.006
	10	Having a furniture refinishing hobbyist in the home	0.006
	11	Living in an old home (more than 10 years)	0.007
	12	Furniture refinishing	0.02
	13	Aged between 40 and 65	0.03
1,1,1-Trichloro-ethane	1	Wood processing	0.0001
	2	Employed	0.0001
	3	Working at a textile plant	0.0007
	4	Metalworking	0.006
	5	Having a metal worker in the home	0.008
	6	Having a chemical worker in the home	0.009

Table 9. (Continued).

Chemical	Rank	Activity[a]	p[b]
Trichloro-ethylene	1	Wood processing	0.002
	2	Working at a plastics plant	0.003
	3	Gas furnace	0.01
	4	Working at a scientific lab	0.01
	5	Smoking	0.02
Tetrachloro-ethylene	1	Employed	0.0001
	2	Wood processing	0.0002
	3	Visiting a dry cleaners	0.003
	4	Working at a textile plant	0.01
	5	Using pesticides	0.01
	6	Working at/being in a paint store	0.03
	7	Being male	0.04
Carbon tetra-chloride	1	Aged less than 17	0.0005
	2	Metalworking	0.006
	3	Working at/being in a paint store	0.02
	4	Furniture refinishing	0.03
m,p-Dichloro-benzene	1	Working at a hospital	0.0001
	2	Having central air conditioning	0.004
	3	Treating home with pesticides	0.05
Chloroform	1	Working at/being in a paint store	0.007
	2	Using pesticides	0.02

[a]Based on questionnaire data from 352 subjects in Bayonne-Elizabeth, New Jersey--Fall 1981.
[b]Probability that the association is due to chance.

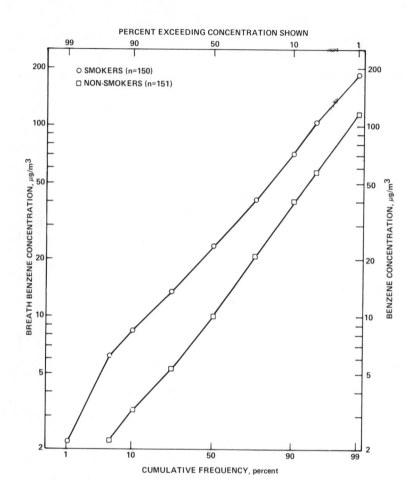

Figure 2. Frequency distributions of benzene levels in breath
of smokers and nonsmokers, measured in the fall of
1981 in Bayonne and Elizabeth, NJ. Smokers had about
twice as much benzene in their breath as nonsmokers.

with the presence of indoor sources in homes that are progres-
sively less open as winter approaches.

These findings are supported by recent studies in Europe
and the United States, some using different adsorbents than
Tenax. Lebret [11], using activated charcoal, found that 35 of
35 organics displayed mean indoor-outdoor ratios greater than 1
in 134 Netherlands homes. De Bortoli [12] found that 32 of 32
organics had indoor-outdoor ratios greater than 1.

It seems clear that many indoor sources of toxic organics exist; however, few have been unequivocally identified and fewer still have had their source emission rates estimated [13]. Identification of indoor sources from among thousands of consumer products and building materials is required to allow a better estimate of possible risks to public health and corrective actions that can be taken.

Two particularly clear examples of indoor chemicals were paradichlorobenzene, used in moth crystals and deodorants; and styrene, used in plastics, foam rubber, and insulation. Tetrachloroethylene (and sometimes 1,1,1-trichloroethane) is used in dry cleaning. Paints may contain vinylidene chloride, styrene, and xylenes. Gasoline contains benzene, ethylbenzene, and xylenes. Tap water contains chloroform, and heated water (particularly hot showers) will give up most of its chloroform to the indoor air [14]. Cleansers also may be sources of chloroform [15]. Benzene was more prevalent in smokers' homes than in nonsmokers'; and smokers' breath levels were about double those of nonsmokers (mean value of 33.5 ± 2.6 (S.E.) $\mu g/m^3$ vs. 16.7 ± 1.5 $\mu g/m^3$).

Outdoor Air

Reliance on outdoor monitors to estimate exposure is contraindicated by this study. Correlations with personal exposures were poor. Median and maximum personal exposures were always greater, sometimes 20 to 50 times greater, than outdoor concentrations, whether measured in this study or in other U.S. cities [16].

Drinking Water

Drinking water was a main source of exposure for 2 trihalomethanes: chloroform and bromodichloromethane. Assuming one l/day cold water intake and 10 m^3/day air intake, the weighted arithmetic mean daily intake of chloroform was 70 μg through water and 90 μg through air (fall 1981). However, for 10 other prevalent chemicals, drinking water usually supplied less than 1% of the total daily intake.

Breath

Breath is an important mode of intake and excretion for many volatile compounds [17]. Whatever compounds are measured in the exhaled breath of a person breathing pure air have been supplied by the bloodstream as it passes through the lungs. The advantages of measuring breath rather than blood are (1) the technique is noninvasive and therefore preferable for use in studies requiring reasonable response rates from general public volunteers; and (2) the measurement technique employed (Tenax; GC-MS analysis) is more sensitive than the corresponding technique for blood employed in the first phase of the TEAM Study. One disadvantage is that only recent exposures are reflected in the breath.

Simple comparisons of exposure to breath concentrations do not take into account the dependence of breath levels on preexisting concentrations in the body and also on the effective biological residence times of each chemical. A simple two-parameter time-dependent model has been developed that accounts for the effect of the initial breath concentration and the effective residence time in the body [18]. The model predicts an effective half-life of 21 hours for tetrachloroethylene and 9 hours for 1,1,1-trichloroethane. A later "washout" study [19] performed over a 10-hour period in a pure air chamber on an adult male exposed for 1 hour to tetrachloroethyelene vapors in a dry cleaning shop interior resulted in a measured effective half-life of 21 hours.

Breath concentrations were significantly correlated with personal exposures to 10 prevalent compounds. Thus, the feasibility of using breath measurements to estimate recent or continuous exposure to these compounds has been demonstrated. For example, breath measurements of persons living near hazardous waste sites could be used to detect current or recent exposure.

ACKNOWLEDGMENTS

Local and state officials in New Jersey gave essential support to this study. Special efforts were made by Dr. John Sakowski and Mr. David Roach of the Bayonne Department of Health, Mr. John Surmay and Mr. Robert Travisano of the Elizabeth Health, Welfare and Housing Department, and Dr. Thomas Burke of the New Jersey Department of Environmental Protection. We thank Sandy Baucom and Shirley Barbour for deciphering the authors' hieroglyphics and creating readable typescripts throughout many revisions. One of us is grateful to Dr. John Spengler and the Harvard University School of Public Health for providing an atmosphere conducive to study. We are most indebted to the hundreds of citizens who conscientiously wore monitors, kept diaries, and answered questions about their activities. The opinions expressed are those of the authors and do not reflect official positions of the U.S. Environmental Protection Agency.

A portion of the material presented in this chapter appeared in "Personal Exposures, Indoor-Outdoor Relationships, and Breath Levels of Toxic Pollutants Measured for 355 Persons in New Jersey," by L. Wallace, E. Pellizzari, T. Hartwell, C.

Sparacino, L. Sheldon and H. Zelon, which first appeared in Atmospheric Environment, 19:1651-1661. (Pergamon Press, Ltd.,
1985).

REFERENCES

1. Wallace, L. A., R. Zweidinger, M. Erickson, S. Cooper, D.
   Whitaker and E. D. Pellizzari. "Monitoring Individual
   Exposure: Measurements of Volatile Organic Compounds in
   Breathing-Zone Air, Drinking Water and Exhaled Breath,"
   Environment International 8:269-282 (1982).

2. Entz, R., K. Thomas and G. Diachenko. "Determination of
   Volatile Halocarbons in Food by Headspace Analysis," J.
   Agric. Food Chem. 30:846-849 (1982).

3. Wallace, L. A., E. Pellizzari, T. Hartwell, M. Rosenzweig,
   M. Erickson, C. Sparacino and H. Zelon. "Personal Expo-
   sure to Volatile Organic Compounds: Direct Measurement in
   Breathing-Zone Air, Drinking Water, Food, and Exhaled
   Breath," Environmental Research 35:293-319 (1984).

4. Wallace, L. "Total Exposure Assessment Methodology (TEAM)
   Study: Summary and Analysis, Volume I," Final Report, Contract No. 68-02-3679, U.S. EPA (1986).

5. Pellizzari, E. D., K. Perritt, T. D. Hartwell, L. C.
   Michael, R. Whitmore, R. W. Handy, D. Smith and H. Zelon.
   "Total Exposure Assessment Methodology (TEAM) Study:
   Elizabeth and Bayonne, New Jersey; Devils Lake, North
   Dakota; and Greensboro, North Carolina: Volume II," Final
   Report, Contract No. 68-02-3679, U.S. EPA (1986).

6. Wallace, L., E. Pellizzari, T. Hartwell, C. Sparacino, L.
   Sheldon and H. Zelon. "Personal Exposures, Indoor-Outdoor
   Relationships, and Breath Levels of Toxic Air Pollutants
   Measured for 355 Persons in New Jersey," Atmos. Env.
   19:1651-1661 (1985).

7. Krost, K. J., E. D. Pellizzari, S. G. Walburn and S. A.
   Hubbard. "Collection and Analysis of Hazardous Organic
   Emissions," Anal. Chem. 54:810 (1982).

8. Bellar, T. A. and J. Lichtenberg. "Determining Volatile
   Organics at Microgram-per-Litre Levels by Gas Chromatography," J. Amer. Water Assoc. 66:739-744 (1974).

9. Conover, W. J. Practical Nonparametric Statistics, 2nd
   ed. (New York: John Wiley, 1980).

10. Evans, J. S., D. W. Cooper and P. Kinney. "On the Propagation of Error in Air Pollution Measurements," Env. Mon. and Assess. 4:139-153 (1984).

11. Lebret, E., H. J. Van de Wiel, H. P. Bos, D. Noij and J. S. M. Boleij. "Volatile Hydrocarbons in Dutch Homes," in Indoor Air, v. 4, pp. 169-174, Swedish Council for Building Research, Stockholm, Sweden (1984).

12. De Bortoli, M., H. Knoppel, E. Pecchio, A. Peil, L. Rogora, H. Schauenberg, H. Schlitt and H. Vissers. "Integrating 'Real Life' Measurements of Organic Pollution in Indoor and Outdoor Air of Homes in Northern Italy," in Indoor Air, v. 4, pp. 21-26, Swedish Council for Building Research, Stockholm, Sweden (1984).

13. Girman, J. R., A. T. Hodgson and A. S. Newton. "Volatile Organic Emissions from Adhesives with Indoor Applications," in Indoor Air, v. 4, pp. 271-276, Swedish Council for Building Research, Stockholm, Sweden (1984).

14. Bauer, U. "Human Exposure to Environmental Chemicals - Investigations of Volatile Organic Halogenated Compounds in Water, Air, Food, and Human Tissue," (text in German), Zbl. Bakt. Hyg., I. Abt. Orig. B., 174:200-237 (1981).

15. Wallace, L., E. Pellizzari, B. Leaderer, T. Hartwell, K. Perritt, H. Zelon and L. Sheldon. "Assessing Sources of Volatile Organic Compounds in Homes, Building Materials, and Consumer Products," paper presented at Conference on Characterization of Sources of Indoor Air Contaminants, Raleigh, NC, May 13-15, 1985; (in press, Atmos. Env.).

16. Brodzinsky, R. and H. Singh. "Volatile Organic Chemicals in the Atmosphere: An Assessment of Available Data," Environmental Sciences Research Laboratory, U.S. Environmental Protection Agency, Research Triangle Park, NC (1982).

17. Krotoszynski, B. K., B. D. Gabriel, H. J. O'Neill, and M. P. A. Claudio. "Characterization of Human Expired Air: A Promising Investigative and Diagnostic Technique," J. Chromatog Sci. 15:239-244 (1977).

18. Wallace, L., E. Pellizzari, T. Hartwell, D. Sparacino and H. Zelon. "Personal Exposure to Volatile Organics and Other Compounds Indoors and Outdoors - the TEAM Study," paper #83.912 presented at the 76th Annual National Conference of the Air Pollution Control Association, Atlanta, GA, June (1983).

19. Gordon, S., L. Wallace and E. Pellizzari. "Breath Measurements in a Clean-Air Chamber to Determine 'Wash-out' Times for Volatile Organic Compounds at Normal Environmental Concentrations," submitted for publication (1986).

CHAPTER 16

INHALATION EXPOSURES IN INDOOR AIR
TO TRICHLOROETHYLENE FROM SHOWER WATER

Julian B. Andelman, Amy Couch, and William W. Thurston

INTRODUCTION

The principal interest in the possible health impacts from air exposure to volatile constituents from the domestic use of water has centered on radon-222 and the short-lived daughters. Prichard and Gesell [1] concluded from their studies that such exposures in the general population of Houston, Texas, alone can produce annual total population doses of the same magnitude as that to the United States population as a result of these radioisotopes mobilized by the mining and milling of uranium. They stated that "radon carried by groundwaters might be an important source of population exposure nationwide."

Prichard and Gesell showed that typically, 50% of the radon is transferred from water to air from all indoor water uses, but the transfer efficiency is highly dependent on the use [1]. The range is 30-90%, the value for showers being 63%. Using a simple one-compartment indoor air model with quantities of water uses similar to those of Prichard and Gesell, and assuming that volatilization is complete, we predicted a linear relationship between incoming water concentration, C_W, and resulting average indoor air concentration, C_A [2]. The relationship is $C_A = 0.0006C_W$, with both concentration terms having the same units, e.g., mg/m^3. If only 50% were to volatilize, the proportionality constant would be reduced similarly by a factor of 0.5.

This relationship was used to compare the likely average human exposures by inhalation and ingestion [2]. For an adult, it was assumed that the person remained in the home all day and the volume of air breathed was 20 m^3. Assuming that two liters of water were ingested, it was calculated that the inhalation exposure would be six times that from ingestion, indicating that indoor inhalation exposures to volatilizing chemicals from all water uses can be substantial and possibly greater than those from ingestion.

We investigated the possible volatilization of trichloroethylene (TCE) into indoor air in buildings in a small community using individual wells obtaining water from an aquifer measured to contain about 40 mg TCE/l [2]. Prior to turning on water in bathrooms, no TCE could be detected in the indoor air above the detection limit for the instrument, namely 0.5 mg/m^3. However, TCE was detected readily in the bathrooms with water running. The air concentration levels increased with time, as expected. In one home, the highest concentration measured after 17 minutes of the shower running was 81 mg/m^3.

It is likely then that showers and baths could constitute important inhalation exposures within the home to volatilizing organic contaminants like TCE. This could result in both a point source of exposure to the person in the bathroom, as well as to inhabitants elsewhere, as the bathroom air is disseminated throughout the home. To investigate further the factors that influence TCE air concentrations resulting from showers, a scaled-down model shower was constructed and operated with known concentrations of TCE injected continuously into the inlet water in the range of 1.5-2.9 mg TCE/l. The results of these experiments were reported briefly elsewhere [2]. This paper will provide more details of the experiments and the mathematical mass-balance models for the change in air concentration as a function of time in the experimental shower system. The ultimate goal is to predict the inhalation exposures that can occur.

EXPERIMENTAL

The experimental shower chamber was a 100 liter glass aquarium standing on its end, the side opening being covered by a rigid Lucite sheet. This cover had holes of various sizes for placing the shower head into the chamber, draining the water effluent, and air sampling and ventilation. The shower head, 1.5 cm in diameter, had six holes drilled in a symmetrical pattern, each being 0.05 cm in diameter. The water was pumped through rigid Teflon tubing to the shower head, the height of which within the shower chamber was adjustable. The locations of the air outlets and inlets within the chamber were also adjustable by the placement of the ends of the tubing. Plastic tubing was passed through a side port to the bottom of the chamber to pump out and/or sample the drain effluent.

Distilled water was pumped from a stainless steel reservoir at a controlled flow rate and mixed with an aqueous concentrate of TCE (trichloroethylene) prior to delivery to the shower head. The blending rate and final shower head concentration was achieved by delivering the TCE concentrate at a controlled rate with a syringe injection pump (Sage Instruments Model 341). For the 43° C experiments the water reservoir was heated.

The chamber air was analyzed by pumping it to one of two real-time continuous monitoring instruments. For most of the experiments an infra-red detection system was used, the MIRAN™ 1A general purpose gas analyzer (Wilks Infra-red Center, Foxboro Analytical). This is a single beam, variable wavelength spectrometer equipped with a gas cell with a path length adjustable from 0.75 to 20.25 m. For some experiments an organic vapor analyzer equipped with a flame ionization detector was used (Century Systems Corp. Model OVA-118). Drain-water samples were analyzed using a purge-and-trap concentration system (Tekmar Company Model LSC-1) in conjunction with a gas chromatograph using a flame ionization detector (Perkin-Elmer Model Sigma II).

The shower experiment was conducted by pumping the aqueous TCE solution through the shower head, typically for approximately one hour, while monitoring the air continuously and for some experiments sampling the drain water for subsequent analysis. After the TCE injection was stopped, water was still pumped through the shower head while the decaying air concentration of TCE was being monitored.

The characteristics of the experimental system are summarized in Table 1. The principal variables that were investigated were TCE concentration, water temperature, and height of the shower drop path.

Table 1. System Variables for Experimental Shower.

Characteristics	Magnitude
Chamber volume, V_A	$0.1 \ m^3$
Air flow rate, F_A	$0.0054 \ m^3/min$
Water flow rate, F_W	$0.00028 \ m^3/min$
TCE initial water conc., C_W	$1.5-2.9 \ g/m^3$
Water temperature, T	23° C, 4° C
TCE injection period	55-60 min
Shower drop path	0.025-0.1 m

RESULTS AND DISCUSSION

Typical results of shower volatilization experiments are shown in Figures 1, 2, and 3. The TCE concentration measured in the effluent air pumped from the shower chamber is plotted as a function of time. As expected, during the 55-60 minute

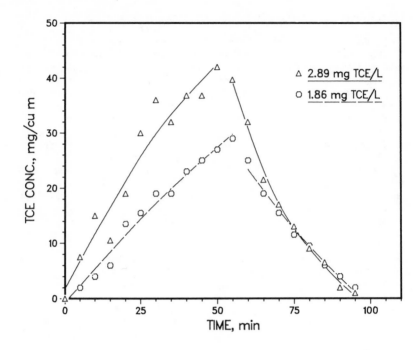

Figure 1. Effect of TCE water concentration on TCE air concen-
tration in model shower. Reprinted from "Inhalation
Exposure in the Home to Volatile Organic Contaminants
of Drinking Water" by Julian B. Andelman in Science
Total Environ., Vol. 47, 1985. Copyright 1985
Elsevier Science Publishers B.V.

injection period the TCE air concentration increased as it volatilized within the shower chamber. After the TCE injection
was terminated, but with shower water still flowing into the
chamber, the TCE air concentration decayed as air also continued to be pumped through the system.
     It should be emphasized that the air flow through the show-
er chamber to the continuous air monitoring system achieves the
important purpose of providing air movement that might occur in
an actual domestic bathroom, although not necessarily at the
precise scaled-down rate. Referring to the system variables
shown in Table 1, a calculation indicates that the air flow
rate relative to the chamber volume is 0.054 chamber volumes/
min, or 3.2/hr. Although this is high compared to perhaps 1/hr
in a home, such air movement does have a controlling influence
on the concentrations obtained due to volatilization. The impact of this variable will be part of the mass-balance treatment in the subsequent discussion.

k_{max} would be equal to F_W (complete volatilization). Using $H = 0.5$ for TCE, we see that for an F_A value of 5.4 1/min, the maximum value of k/H would be 0.56 1/min (0.28 1/min divided by 0.5). It is thus apparent that, as a good approximation in this system, k/H can be essentially neglected compared to F_A and Equation 9 would simplify to

$$V_A(dC_A/dt) = kC_W - F_AC_A. \qquad (10)$$

Equation 10 has the same form as Equation 6 and implies that for TCE in this shower system one can essentially assume that the effective rate of volatilization, kC_W, is constant.

A different approach can be taken effectively to reach this same conclusion by examining the experimental results shown in Figure 1. With an injected concentration of 2.89 mg TCE/1, the maximum measured air concentration was approximately 40 mg/m^3. Assuming that as much as 80% of the TCE was volatilizing at this point (see Table 2), 0.56 mg TCE/1 would remain in the drain water, or 560 mg TCE/m^3. The concentration of TCE in the water diffusion film at the air side of the interface, C_{WF}, is equal to C_A/H, or in this case 80 mg TCE/m^3 (40/0.5). Using these values in Equation 7, it is apparent that C_W is substantially larger than C_A/H in this system, and again one can conclude that as a good approximation $R = kC_W$, the volatilization rate being essentially constant.

One can estimate the steady-state concentration in the air that could be attained. Using Equation 10, at steady-state dC_A/dt equals zero and rearrangement gives

$$C_A(\text{steady-state}) = kC_W/F_A. \qquad (11)$$

Thus, the steady-state air concentration is directly proportional to the incoming water concentration, C_W, and inversely so to the air flow rate. Equation 11, based on a constant volatilization rate model, thus predicts a steady-state air concentration relationship of the same form as that of Equation 5, which is based on the Henry's Law equilibrium. However, the proportionality constant of Equation 5 cannot be exceeded by that of Equation 11, which in any event is a simplification, as discussed above. In some of the experiments it appears that steady-state was almost attained, such as in Figure 2, the 23° C curve. In such a case, k can be estimated using Equation 11, since C_W and F_A are known.

The air concentration buildup as a function of time can also be derived by integrating Equation 10 to obtain

$$\ln(1 - F_AC_A/kC_W) = -(F_A/V_A)t. \qquad (12)$$

This equation models the buildup portion of the shower volatilization curves shown in Figures 1, 2, and 3. Where k is known for a system approaching steady-state, one can test to see if Equation 12 accurately describes the behavior of the system by plotting log $(1 - F_AC_A/kC_W)$ as a function of time. This is the same function as log $(1 - C_A/C_A(\text{steady-state}))$. If

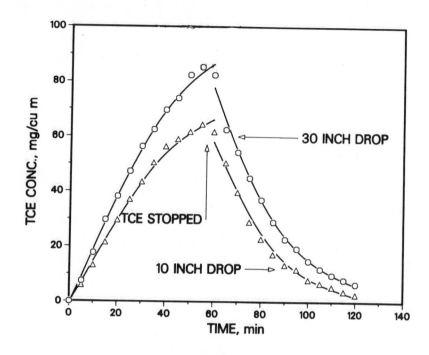

Figure 3. Effect of height of drop path on TCE air concentration in model shower [2].

In the shower system with air and water flowing continuously,

$$V_A/V_W = F_A/F_W = 5.4/0.28 = 19.3.$$ (2)

Combining Equations 1 and 2 for this system and using the definition $H = C_A/C_W$, it follows that

$$M_A/M_W = 19.3 H.$$ (3)

In these calculations, the dimensionless form of H is being
used. Since the mass ratio for the volatilized constituent in
the shower system is a simple function of H, as shown in Equation 3, so therefore is the fraction volatilized, f, which is
expressed as $M_A/(M_A + M_W)$. Using the latter in conjunction with Equation 3, one obtains

$$f = 1/(1 + 1/[19.3 H]).$$ (4)

Similarly, one can express the predicted equilibrium or maximum
air concentration, C_A, as a function of H and the <u>initial</u>
water concentration, C_{Wi}, using the Henry's Law <u>definition</u>

and the fact that C_W at equilibrium would be $C_{Wi}(1 - f)$, giving

$$C_A = H(1 - f)C_{Wi}. \qquad (5)$$

Table 3 shows these relationships for TCE and, for comparison, chloroform in the experimental shower system. Obviously, the higher H value for TCE compared to chloroform (a factor of four) predicts that its equilibrium air/water mass ratio is similarly higher than that for chloroform by a factor of four. Although 91% of the TCE is predicted to volatilize if equilibrium were attained, a maximum of only 71% of the chloroform would similarly be expected to volatilize because of its smaller H value. Similarly, the expected equilibrium air concentrations for these two chemicals are simply predicted from the initial water concentrations, C_{Wi}, the multiplying factor for TCE, 0.045, being higher than that for chloroform by the ratio of their f values.

Table 3. Maximum Equilibrium Volatilization in Shower System as Affected by H.

Chemical	H	M_A/M_W	f	C_A
TCE	0.5	9.65	0.91	0.045 C_{Wi}
Chloroform	0.125	2.41	0.71	0.036 C_{Wi}

As shown in Table 3, the actual extent of volatilization for TCE in the shower system ranged from 43% to 67% at room temperature, although as expected, considerably more volatilized at the higher temperature of 41° C. The room temperature volatilization was substantially less than the 91% prediction shown in Table 3 if Henry's Law equilibrium were attained. This is a good indication that mass transfer controlled and limited the rate of volatilization.

As discussed elsewhere [2], kinetic-mass-balance relationships for a chemical volatilizing into the shower system can be developed to describe the change in air concentration as a function of time within the shower chamber. As the shower water flows into the chamber and the TCE (or other chemical) volatilizes, air is passed simultaneously through the chamber (as can occur in an actual domestic shower). The discussion above deals with the steady-state attainment of air concentrations on the assumption that Henry's Law equilibrium is attained. However, time transients are of interest, particularly if such equilibrium cannot be reached readily because of mass-transport limitations. The changes in concentration as a function of time shown in Figures 1, 2, and 3 should be explicable in terms of mass-balance and the rate of volatilization.

Taking the air volume of the shower chamber as V_A and the rate of volatilization as R (mass of chemical volatilized per unit time), the mass balance equation for the rate of change of TCE chamber air concentration at any time can be expressed as

$$V_A(dC_A/dt) = R - F_A C_A. \tag{6}$$

In the earlier paper [2], this equation was integrated to obtain relationships for the buildup in air concentration as a function of time, the steady-state value, C_A(steady state), and the decaying concentration once the TCE was no longer being injected, but air was still being passed through the chamber. Before discussing these further, it should be noted that mass-transfer rates for chemicals volatilizing across a water-air interface are often modeled in terms of the driving force being a concentration gradient across a diffusion-limiting liquid (water) film. The difference in concentration between the bulk water at the solution side of the interface and that at the air-water side establishes the diffusion concentration gradient [3]. Immediately upon volatilization of the TCE from the shower water an air concentration is established, C_A, adjacent to the air-water interface. Thus the concentration in the water, C_{WF}, in the diffusion liquid film on the air side, can be taken to be in equilibrium with that in the air. It can then be expressed as $C_{WF} = C_A/H$. On this basis, the R value in Equation 6 is

$$R = k(C_W - C_A/H), \tag{7}$$

where k is the volatilization transfer coefficient with units of volume per time (e.g., m^3/min). It is thus apparent that the rate of volatilization itself may not be constant if the C_A/H term in Equation 7 is substantial compared to C_W and the concentration of volatilized chemical in the air of the chamber builds up with time.

Using Equation 7 in conjunction with 6, one obtains

$$V_A(dC_A/dt) = k(C_W - C_A/H) - F_A C_A. \tag{8}$$

On rearrangement this takes the form

$$V_A(dC_A/dt) = kC_W - C_A(F_A + k/H). \tag{9}$$

This equation can be simplified further by comparing the relative magnitudes of the F_A and k/H terms. An upper limit for the value of k is simply F_W, the flow rate of the shower water. This can be seen by considering the definition of R in Equation 7. Initially, before any significant magnitude of C_A has been attained, the maximum volatilization rate that could occur would be $F_W C_W$, namely all of the TCE. Thus

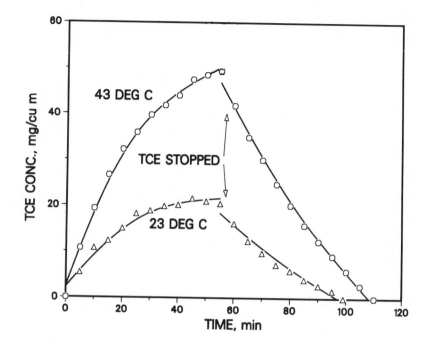

Figure 2. Effect of water temperature on TCE air concentration
in model shower [2]. Reprinted from "Inhalation Exposure in the Home to Volatile Organic Contaminants
of Drinking Water" by Julian B. Andelman in Science
Total Environ., Vol. 47, 1985. Copyright 1985 Elsevier Science Publishers B.V.

Figure 1 shows the expected higher air TCE concentrations
in the experiment with the higher concentration of TCE in the
injected water. Similarly, Figure 2 shows increased volatilization at the higher water temperature, while Figure 3 indicates that when the height of the shower water drop path increased, so did the rate of volatilization. It should be
noted, however, that this effect was not always obtained, possibly due to variability among experiments. Nevertheless, it
is consistent with the likelihood of increased volatilization,
as the water droplet is exposed to the air for a longer time
period.

Table 2 represents mass-balance and related factors for
several typical experiments. Mass-balance was assessed by
comparing the quantity of TCE injected in each experiment with
that measured in the air and water effluents. The latter was
determined from periodic collections and analyses of drain
water, while the former was calculated by integrating the area

Table 2. Mass-Balance in Typical Shower Experiments.

Experiment Type	Total TCE Injected, mg	TCE Volat. During Build-up, mg	Percent Volatilized	Total TCE Measured, mg	Percent Recovery
Low conc.	13.3	9.0	67	9.7	73
Normal	19.5	8.3	43	18.3	94
Normal	19.5	11.8	61	21.3	109
Low height	19.5	8.6	44	20.0	103
High temp.	18.1	14.3	79	20.7	114

under the volatilization curves, such as those in Figures 1, 2, and 3. As shown in Table 2, the recovery ranged from 73% to 114%, the variation probably due to analytical imprecision. The volatilization in the buildup period was determined by integrating the buildup portion of the curve and adding the amount remaining in the air of the chamber at the end of this period. In each case then, shown as percent volatilization in Table 2, this represents an integrated average for the total buildup period. Although the absolute magnitude volatilized is greater for the higher injected TCE concentration, it appears that the only clear and substantial effect on percent volatilization is temperature. This is not unexpected, since an increase in temperature will normally increase both the rate of mass transfer across a liquid film, and the Henry's Law constant, H (the equilibrium air concentration of a volatile constituent divided by its aqueous concentration) [2].

As discussed elsewhere [2], one can estimate the maximum volatilization that could occur in the experimental shower system on the assumption that Henry's Law equilibrium is attained. This maximum is determined by H and the relative air and water flow rates through the chamber, F_A and F_W, respectively. For the shower system F_A was 5.4 1/min typically, while F_W was 0.28 1/min. The mass of the volatilizing chemical at equilibrium distributing itself between the air and water phases, M_A/M_W, can be expressed in terms of the equilibrium concentration ratios in these two phases, C_A/C_W, and the volume ratio of air and water phases, V_A/V_W, as follows:

$$M_A/M_W = (C_A/C_W)(V_A/V_W).$$ (1)

one obtains a linear plot and the slope/2.3 equals F_A/V_A, this indicates that the model is at least consistent with the measured volatilization curves. Such a plot is shown in Figure 4. The slope of 0.055/min is almost identical to the value of 0.054/min calculated from the ratio F_A/V_A.

Once the source of the volatilizing chemical is eliminated (injection into the shower terminated), its air concentration should gradually decrease as it is diluted by incoming air. The form of decay is predicted by integrating Equation 10 with the term kC_W equal to zero. One then obtains a typical first-order decay relationship

$$\ln(C_{A1}/C_{A2}) = (F_A/V_A)(t_2 - t_1). \tag{13}$$

A plot of this, shown in Figure 5 for a typical experiment, indicates that the decay is first order with a slope of 0.046/min, again reasonably close to the 0.054 value for F_A/V_A.

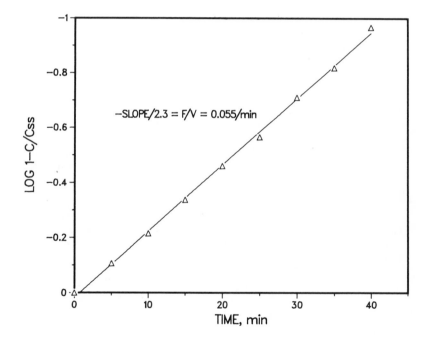

Figure 4. TCE buildup function versus time in model shower system. See Equation 12 and subsequent discussion for explanation of terms. The text uses subscript "A" for the parameters C, Css, F, and V.

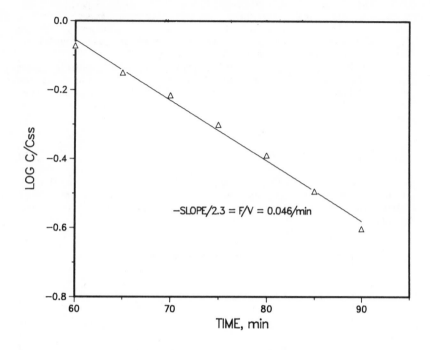

Figure 5. TCE decay function versus time in model shower system. See Equation 12 and subsequent discussion for explanation of terms. The text uses subscript "A" for the parameters C, Css, F, and V.

The above analysis indicates that TCE volatilization in the scaled-down experimental shower system can be modeled and assessed in terms of predictable volatilization and mass-balance consideration. The extent of the volatilization is substantial and greatly affected by temperature. The chamber volume and air flow rate through it also affect the resulting air concentrations, as expected. Thus, the human exposure that can result will clearly be determined by all these factors in a full-scale domestic shower.

Although the time periods studied in these experiments are substantially longer than those that are likely to be encountered in an actual domestic shower, ranging perhaps from 5 to 15 minutes, the data do indicate that in the earlier stages the shower chamber air concentrations of TCE increase approximately in a linear manner with time. This has potentially important implications for human exposures. If the bather were exposed to a constant air concentration, this would imply that the exposure would increase proportionally to the time spent in the shower. However, with the air concentration increasing as a

function of time, the total exposure will increase exponential-
ly. For example, if we take the air concentration, C_A, to be
equal to kt, and exposure is the time-integrated product of the
rate of inhalation times C_A, then it follows that the time-
integrated exposure will increase as the square of the time of
exposure in the shower. This suggests that limiting the period
of shower use can substantially reduce the inhalation exposure
to the user.

Finally, one can conclude that TCE volatilizing in a shower
may constitute a significant point source of human exposure for
the bather, and a dispersed source for others elsewhere in the
home. However, other indoor water uses should also be considered from the point of view of possible inhalation exposures,
and all possible inhalation exposures from indoor water uses
should be compared to exposures from the direct ingestion of
such contaminated water. For highly volatile chemicals, these
inhalation exposures have the potential for being substantially
greater than those associated with the direct ingestion of
water.

ACKNOWLEDGMENTS

This research has been funded in part by the U.S. Environmental Protection Agency (EPA) under assistance agreement CR
811173-01 with the Center for Environmental Epidemiology, Graduate School of Public Health, University of Pittsburgh. The
authors gratefully acknowledge this support and in particular
the encouragement and advice of the EPA project officer,
Gunther Craun, at the Cincinnati Health Effects Research Laboratory.

REFERENCES

1. Prichard, H. M., and T. F. Gesell. "An Estimate of Population Exposures Due to Radon in Public Water Supplies in the
   Area of Houston, Texas," Health Physics 41:599-606 (1981).

2. Andelman, J. B. "Inhalation Exposure in the Home to Volatile Organic Contaminants of Drinking Water," Science Total
   Environ. 47:443-460 (1985).

3. Mackay, D., and A. T. K. Yeun. "Mass Transfer Coefficient
   Correlations for Volatilization of Organic Solutes from
   Water," Environ. Science Technol. 17:211-217 (1983).

CHAPTER 17

DRINKING WATER CHARACTERISTICS AND CARDIOVASCULAR
DISEASE IN A COHORT OF WISCONSIN FARMERS

Elaine A. Zeighami, Gunther F. Craun, and Charlotte A. Cottrill

INTRODUCTION

In 1957, Kobayashi first reported a statistical correlation
between the acidity of water supplies in Japan and cerebrovascular mortality [1]. Since then, numerous studies have reported an association between drinking water quality and cardiovascular disease [2-13]. Several excellent reviews of these
studies have recently been published [2-8]. In general, studies have shown an inverse association between water hardness
and cardiovascular disease mortality; that is, lower mortality
has been found in areas where the water hardness is high. Most
of the studies reporting this association were descriptive or
ecologic epidemiologic studies of mortality rates in broad geographic areas having different water characteristics. Only a
few studies have considered tap water exposures and possible
confounding by various risk factors or have provided an estimate to measure the possible effect of a water factor [10-13].
This can be accomplished through analytic epidemiologic studies.
In an analytic epidemiologic study, information on exposure
and disease is available for each individual, and a quantitative measure of the association is obtained. Appropriate study
designs include the prospective cohort, retrospective cohort,
case-control, and cross-sectional epidemiologic studies. Although analytic epidemiologic studies are more costly and difficult to conduct than descriptive or ecologic studies, they
offer information about causal interpretations. It is particularly important in analytic studies to accurately assess and
define exposure and disease status to avoid random misclassification which will result in decreased study sensitivity.

Random misclassification can only bias the study toward observing no association between exposure and disease.

In studies of drinking water associations with cardiovascular disease, accurate information must be obtained not only for each subject's exposures to the drinking water constituents of interest, but also for other exposures and other risk factors, as the data relating exposure to disease may convey an appearance of association because of confounding bias. Although negative confounding can also occur, the primary concern is that confounding has led to the erroneous observation of an association. Confounding bias is a basic characteristic of any epidemiologic study, and does not necessarily result from any error on the part of the investigator. Information should be collected on known or suspected confounding characteristics. If a characteristic can be demonstrated to have no association with either the exposure or disease being studied, that characteristic cannot be confounding. To prevent confounding, matching is generally employed in the study design. To assess and control confounding during data analysis, stratification or multivariate techniques are employed.

Some of the specific design considerations for drinking water studies include the following: the degree of uniformity of exposure within a community having a common public water supply; exposure to numerous constituents in drinking water; correlations of certain water contaminants; potentially wide ranges of concentrations for certain water constituents; the concentrations of many water contaminants which are undetectable by current analytic techniques; and additional exposure to similar constituents in food.

In an attempt to minimize some of these methodological difficulties in designing an analytic epidemiologic study of the association between drinking water quality and cardiovascular disease, we chose to conduct a case-control study within a large cohort having individual well water supplies. The cohort resided in Wisconsin, a state whose ground water displays a range of hardness levels, a feature that allowed a variety of water hardness levels to be included in the study.

This cohort, which consisted of farmers, was selected because it was a large, relatively homogeneous population of males, virtually all of whom had individual drinking water supplies. Furthermore, the persons in this group had only one primary drinking water supply, unlike most other employed groups who may have had different water supplies at home and in the work place. The average age of the population was over 50 years [14], so the population was primarily in the high-risk group for cardiovascular disease events.

SELECTION OF CASES AND CONTROLS

Cases were ascertained from death certificates in which at least one cause of death listed was coronary artery disease

(ICDA 410-414) or cerebrovascular disease (ICDA 430-438) and the occupation listed was farmer. The next-of-kin was contacted by mail two to three months after the date of death and invited to participate in the study. The mailing included a water sampling kit and a questionnaire. The respondent was asked to verify that farming was the deceased's primary lifetime occupation and that he resided on a farm for two years prior to his death.

For each non-coroner-certified death identified through death certificate screening, the certifying physician was contacted by mail and asked to provide more detailed information on cardiovascular disease and causes of death. Excluded from the study were cases for whom the physician did not verify cause of death as coronary artery disease (CAD) or cerebrovascular disease (CBVD). However, the cases for whom the certifying physician did not return a questionnaire were retained in the study and the causes of death were presumed to be those on the death certificates. Of the non-coroner-certified cases, 259 (66%) had physician verification of causes of death and a cardiovascular disease history, as shown in Table 1. All causes of death were coded by a nosologist.

Table 1. Distribution of Cases.

Physician-verified Diagnoses	Number
Myocardial infarction	148
Cerebrovascular accident	69
Other coronary artery disease	42

Death-certified Information Only	Number

Coroner-certified
  Sudden death, listed as myocardial infarction
  or heart attack
	109

Non-coroner certified Myocardial infarction (ICDA[a] 410)	60
Cerebrovascular disease (ICDA 430-438)	48
Other coronary artery disease (ICDA 411-414)	28

[a]International Cause of Death, 9th Edition.

Controls were obtained from the Brucellosis Testing List, which is maintained and updated annually by the state of Wisconsin for all farms which sell Grade A milk. Controls were selected by stratified random sampling to represent the distribution by county of all farms in the state with sales of $2,500 or more in 1974. Only farms on which a white male at least 35 years of age resided were included in the study. Living controls were chosen because a high proportion of deaths among white males over 35 include a diagnosis of either coronary artery disease or cerebrovascular disease, although they are not necessarily listed as the underlying cause of death. Next-of-kin for both cases and controls were contacted by the same mail procedures.

Only persons who had resided on a farm for at least the previous two years and had not been employed off the farm more than 40% of that time were included in the study. Less than 5% of cases and controls reported any work off the farm. The case group and control group differed considerably in age. Ninety-five percent (95%) of the cases were over 54 years old compared to 44% of the controls. It was impractical to match cases and controls on age because information about the age of controls was not available until they had been contacted. Age was thus included as an independent variable in the multivariate analysis.

Differences in sources for the identification and selection of cases and controls introduce the possibility of fundamental differences between the two groups which may introduce bias. In particular, we were concerned that a higher percentage of dairy farmers might be present in the control population, since they were selected from the Wisconsin Brucellosis Testing List. This listing does not exclude livestock farmers nor does it include dairy farmers who no longer sell Grade A milk. Since 63% of Wisconsin farms are classified as principally dairy, and 77% are either dairy or livestock [14], it is likely that both the case and control series were primarily comprised of individuals who were or had been dairy or livestock farmers.

Respondent questionnaires were completed by a telephone interview with the next-of-kin (spouse) for cases and controls. When a telephone interview could not be arranged, the respondent was asked to complete the questionnaire and return it by mail. The questionnaire included questions on occupation, place and length of residence, nonfarm employment, smoking, diet (principally intake of fatty foods), liquid intake, brief medical history, water sources, use of water-softening equipment, and type of farm.

GEOGRAPHIC AND DRINKING WATER SAMPLING CONSIDERATIONS

A source of potential error in exposure assessment is the use of a single water sample to estimate both current and past content of drinking water. All supplies in this study were ground water supplies, thus increasing the likelihood that the water content of constituents (e.g., calcium and magnesium) arising principally from geochemcial sources is relatively constant. Even though the levels of metals introduced into water through contact with pipes and water storage/pressure tanks could be more variable, a single tap water sample collected on first draw in the morning may still provide a reasonable estimate of the relative levels of these constituents when the water is not corrosive, or when plumbing has not been significantly altered over the years. In general, Wisconsin ground waters are not corrosive. In this study, nitrates seem the constituent most likely to be highly variable. Nitrate contamination primarily results from the leaching of fertilizer and animal waste into ground water, and secondarily occurs in natural geologic deposits. Nitrate levels may be influenced by recent rainfall, recent fertilizer use, seasonality, and other variant factors. Therefore, it is difficult to use nitrate levels from single samples to estimate long-term exposure. For the remaining constituents, it is felt that the data obtained from a single sample is a reasonable estimate of exposure. Historical information on water quality, however, was not available for these water supplies and this single sample represents a potential source of exposure misclassification which might tend to bias the results toward no association.

The content of drinking water is not independent of geographic location; this must be considered when defining the area for selection of cases and controls. In Wisconsin there are distinct regions in which water is generally hard (above 250 ppm) and others where it is relatively soft (below 80 ppm). Even within small geographic areas, we found considerable variation in hardness of drinking water. Thus it is possible that selecting cases and controls without regard to the geographic location of their farms could result in spurious differences between the two groups for hardness and other water constituents. However, matching cases and controls for location could have potentially obscured real differences in water constituents. A characteristic of matching is that if a factor is matched, that factor cannot be evaluated. In this instance if matching on location resulted in also matching on certain water constituents, those constituents could not be evaluated for an association with disease - a flaw we wished to avoid. Therefore, the optimal solution appeared to be to control for location in the analysis by stratifying for region, and to ascertain that cases and controls were not concentrated in different geographic areas. This was done by subdividing the state into three regions which roughly correspond to South, Central, and North, as shown in Table 2. For all three areas, the case and control groups were similarly located by county, which was the smallest geographic unit available within the area. The proportion of cases and controls located in each of the three areas is nearly identical, and the distribution of cases and controls by county within the area is similar.

Table 2. Distribution of Case and Control Groups, By Area.

Area	Proportion of Case Group in Area	Proportion of Control Group in Area
South	0.62	0.62
Central	0.25	0.24
North	0.13	0.14

First-draw morning water samples were taken from the kitchen cold water tap. Respondents were given standardized instructions for filling two 250 ml polyethylene bottles and one gas-tight glass vial for pH determination. Samples were mailed to the Wisconsin State Hygiene Laboratory in Madison, where all water analyses were carried out. One 250 ml bottle was acidified upon receipt and was used for metal analyses. Parameters analyzed, method of analysis, and analytic detection limits are shown in Table 3.

DATA ANALYSIS

Water Constituents

For some constituents, the levels present in almost all of the water samples were below the analytic detection limits. For all constituents except nickel, tin, and barium, the distribution of the values above detection is approximately lognormal. Constituents which were found in less than 1% of samples were not included in any analysis. A more difficult problem was the handling of constituents which were below analytic detection in a large proportion of samples, but which are important from a biological standpoint, e.g., lead and cadmium. The value of a water constituent below detection must be carefully interpreted, since information is provided that the value of the element is at or below the analytic detection limit. One means of treatment in the analysis is to assign each constituent with a value below detection a value of zero. Alternatively, a value equal to some fraction of the analytic detection limit could be assigned.

In selecting methods of data analysis the joint relation of the drinking water constituents must be considered in relation to case-control status. Multivariate analysis in which all constituents were entered as independent variables was conducted. The basic form of analysis in this particular study is

Table 3. List of Water Constituents, Methods of Preservation, and Detection Limits.

Parameter	Method	Preservation	Detection Limit	Samples Below Detection (%)
Calcium	AAS-flame[a]	HNO$_3$	1.0 mg/l	1.5
Magnesium	AAS-flame	HNO$_3$	1.0 mg/l	3.6
Iron	AAS-flame	HNO$_3$	0.1 mg/l	51.2
Zinc	AAS-flame	HNO$_3$	0.02 mg/l	5.7
Copper	AAS-flame	HNO$_3$	0.05 mg/l	41.5
Barium	AAS-flame	HNO$_3$	0.4 mg/l	c
Lead	AAS-HGA[b]	HNO$_3$	0.003 mg/l	74.8
Manganese	AAS-flame	HNO$_3$	0.04 mg/l	80.5
Tin	AAS-flame	HNO$_3$	1.0 mg/l	c
Sodium	AAS-flame	HNO$_3$	1.0 mg/l	2.2
Potassium	AAS-flame	HNO$_3$	1.0 mg/l	12.1
Chromium	AAS-HGA	HNO$_3$	0.003 mg/l	98.5
Cadmium	AAS-HGA	HNO$_3$	0.0002 mg/l	74.3
Nickel	AAS-HGA	HNO$_3$	0.01 mg/l	c
Fluoride	AAS-HGA	HNO$_3$	0.1 mg/l	12.3
Alkalinity	H$_2$SO$_4$ titration	None	1.0 mg/l CaCO$_3$	0.1
Hardness	Calculation	None	1.0 mg/l CaCO$_3$	0.4
Nitrate	Automated cadmium reduction	None	0.02 mg/l	30.9
pH	Potentiometric	None	N/A	N/A

[a]Atomic absorption spectrophotometry, flame atomizer.
[b]Atomic absorption spectrophotometry, heated graphite furnace atomizer.
[c]Detected in less than ten total samples.

the logistic regression of case-control status on the independent variables. All logistic regressions were fit using the LOGIST procedure of SAS Institute, Inc. [15]. All regressions were carried out with the water constituent values below detection set at three levels; zero, half the detection limit, and equal to the detection limit. In each instance, there was at most a very small effect on the beta coefficients (slope) of the regression and no effect on the p-values.

Alkalinity is a term used by water chemists to denote the total carbonate content of the drinking water. The rank correlation coefficients of all water variables in the study show that alkalinity is closely related to total hardness (r = 0.83). Since total hardness is a calculated measure which includes calcium and magnesium, total hardness is closely correlated with both calcium (r = 0.97) and magnesium (r = 0.97). Calcium and magnesium levels are also generally highly correlated (r = 0.92) in drinking water because they arise from similar geochemical sources.

When more than one of these four variables (alkalinity, hardness, calcium, and magnesium) was included in a logistic regression, the high degree of correlation of the two independent variables generally obscured any observation of a relationship. Accordingly, four separate regression models were used in the analysis. Each model included among the independent variables only one of the four "hardness-related" variables; all other remaining water constituents were included. Model 1 contains total hardness, Model 2 contains alkalinity, Model 3 contains calcium, and Model 4 contains magnesium. Age was correlated only with values for metals in water; the effect of including age in the logistic regression is discussed later.

Another important factor to consider in a study of the association of drinking water constituents and cardiovascular disease is the use of individual home water ion exchange units which soften water by removing calcium and magnesium ions. Because these ions are exchanged for sodium ions, the sodium concentration of the tap water is ordinarily increased. The personal reporting of softener use appeared to be highly unreliable in the present study, based on a comparison of reported use of a water softener for drinking water with the sodium levels found in the water samples. Because natural sodium levels are universally low in Wisconsin ground waters, the sodium level of the water samples was felt to be a more accurate criterion for determining whether an individual water supply had been softened by ion exchange. Use of the level of water hardness provided little information, because some water softeners apparently operated inefficiently, accomplishing only a partial exchange. A comparison of the values of various water constituents in the water supplies with low sodium (Na less than 10 mg/l) and high sodium (Na greater than 40 mg/l), showed lower calcium and magnesium values for both the case and control groups with high sodium water supplies, as shown in Table 4. The cutoff points were chosen because those drinking water samples with sodium levels between 10 mg/l and 40 mg/l were felt to be equivocal with regard to softener use. Alkalinity is slightly higher in the high sodium group, as are levels of copper and zinc. These results tend to confirm that sodium levels are a reliable indicator of home water softener use in this cohort, and analyses were conducted using these sodium groups.

Table 4.   Distribution of Selected Water Constituents in Low
           Sodium and High Sodium Groups.[a]

Controls			
Low Sodium Group[b] (N = 752)		High Sodium Group[c] (N = 102)	
Element	Median value (mg/l)	Element	Median value (mg/l)
Calcium	54.0	Calcium	6.5
Magnesium	30.0	Magnesium	3.0
Alkalinity	238.0	Alkalinity	280.0
Potassium	1.0	Potassium	1.0
Copper	0.08	Copper	0.05
Zinc	0.17	Zinc	0.04

Cases			
Low Sodium Group[b] (N = 458)		High Sodium Group[c] (N = 46)	
Element	Median value (mg/l)	Element	Median value (mg/l)
Calcium	56.0	Calcium	13.0
Magnesium	23.0	Magnesium	12.0
Alkalinity	221.0	Alkalinity	250.0
Potassium	1.0	Potassium	2.0
Copper	0.09	Copper	0.0
Zinc	0.35	Zinc	0.12

[a]Values below detection are treated as zeros in the calculation of the median.
[b]Low Sodium Group = less than 10 mg/l sodium.
[c]High Sodium Group = greater than 40 mg/l sodium.

Cause-of-Death-Categories

Table 5 presents the results of each of the four logistic regression models separately for the two cause-of-death categories, CAD and CBVD. For each analysis by cause of death, there are 387 cases of CAD and 117 cases of CBVD. In each model the hardness-related variable is shown in the table; omitted are the other water variables which are not significant at $p < .10$. As anticipated, age is statistically significant since cases were considerably older than controls. Age was included in each model because it was not related to any of the hardness variables. Regression models which did not include age as an independent variable did not result in any change for the four hardness-related variables. The estimates for some of the metals were different when age was not included in the regression models.

Table 5. Logistic Regression Models of Case Status, By Cause of Death Category.

Coronary Artery Disease Deaths

Independent Variable	Beta Coeff.	p-Value	Independent Variable	Beta Coeff.	p-Value
	Model 1[a]			Model 2[a]	
Age	0.1941	.0001	Age	0.1941	.0001
Hardness	-0.0012	.08	Alkalinity	-0.0022	.01
Sodium	-0.0005	.89	Sodium	0.0022	.58
Potassium	0.0143	.27	Potassium	0.0167	.20

Cerebrovascular Disease Deaths

Independent Variable	Beta Coeff.	p-Value	Independent Variable	Beta Coeff.	p-Value
	Model 1[a]			Model 2[a]	
Age	0.2028	.0001	Age	0.2670	.0001
Hardness	-0.0018	.03	Alkalinity	-0.0026	.06
Sodium	-0.0130	.09	Sodium	0.0167	.006
Potassium	0.0340	.13	Potassium	0.0332	.13

Table 5. (continued)

Coronary Artery Disease Deaths

Independent Variable	Beta Coeff.	p-Value	Independent Variable	Beta Coeff.	p-Value
	Model 3[a]			Model 4[a]	
Age	0.1935	.0001	Age	0.1939	.0001
Calcium	-0.0034	.29	Magnesium	-0.0086	.11
Sodium	-0.0001	.99	Sodium	0.0002	.96
Potassium	0.0116	.37	Potassium	0.0136	.29

Cerebrovascular Disease Deaths

Independent Variable	Beta Coeff.	p-Value	Independent Variable	Beta Coeff.	p-Value
	Model 3[a]			Model 4[a]	
Age	0.2028	.0001	Age	0.2695	.0001
Calcium	-0.0072	.18	Magnesium	-0.0153	.07
Sodium	-0.0129	.04	Sodium	0.0137	.02
Potassium	0.0319	.15	Potassium	0.0330	.14

[a]Cr, Fl, NO_3, Cd, Pb, Fe, Cu, Zn, Mn, pH included in model; not statistically significant at $p < .10$.

One major difference was found between the two cause-of-death categories. For CBVD deaths, an association between water sodium level and case status appears in all the regression models, in addition to an association between each hardness value and case status. For CAD deaths, no association between sodium level and case status was found. The relationship with sodium level is positive, so that CBVD cases had higher water sodium levels than did controls at a given level of all the other independent variables in the model. Sodium is significantly higher among CBVD cases in all four regressions (at $p < .01$ in each model). For CBVD, magnesium has a stronger relationship to case status than does calcium.

The association of sodium level to CBVD disease risk is almost certainly due to a higher proportion of softener users among CBVD cases, since ground waters in Wisconsin contain little or no natural sodium; and may actually represent an association between artificial softening of the drinking water and CBVD. That both sodium and hardness are significant in the logistic regressions for CBVD indicates an independent contribution of each.

Odds ratio estimates for the hardness variables, as well as for sodium for CBVD deaths, are presented in Table 6 as a quantitative measure of the observed associations. These estimates are taken from the logistic regression models containing age and the other water variables which were not statistically significant. Thus, the estimates for the odds ratio are estimates at a fixed value of all other water variables included in the regression. A negative sign indicates that the change in the water variable for the odds ratio given is for a decreased concentration of the constituent. For example, the odds ratio for hardness (Model 1) represents a 12% increase in the relative risk for CAD and a 20% increase in relative risk for CBVD for each 100 mg/l decreased water hardness. In all instances, the increased risk associated with a hardness variable and sodium is small or moderate.

Sodium Level and Water Softener Use

All analyses were carried out separately for the low sodium and high sodium groups because that data showed a difference in the values of certain water constituents between these groups. In each instance, the analysis by sodium group resulted in differences in the nature and interpretation of the association between the water hardness variable and CAD or CBVD. Artificial softening of water was considered as a potential confounder of the drinking water hardness-disease association. If, for example, the elevated sodium content of artificially softened water produced a higher disease risk, then if the total group contained substantial numbers of homes with artificially softened water, the appearance of an association between soft water and disease could be conveyed. Stratification on this potential confounder during data analysis was used to assess if the use of home water softeners affected the observed association.

Logistic regression analysis was conducted separately for the two groups: the group of cases and controls with sodium less than 10 mg/l (low sodium), and the group with sodium levels greater than 40 mg/l (high sodium). The dichotomy has some misclassification of softener users, with the greatest error probably being the inclusion in the high sodium group of a few water supplies having high natural sodium levels. However, the magnitude of this error is probably minimal and would not materially affect results.

Table 6. Odds Ratios for Water Constituents, By Cause of Death
Category.[a]

Variable	Amount of Change in Water Variable (mg/l)	Odds Ratio[a]	90% Confidence Interval
Coronary Artery Disease			
Model 1 - Hardness	-100	1.12	(1.00, 1.26)
Model 2 - Alkalinity	-100	1.24	(1.07, 1.44)
Model 3 - Calcium	-50	1.19	(0.91, 1.54)
Model 4 - Magnesium	-20	1.19	(0.99, 1.42)

Cerebrovascular Disease			
Variable	Amount of Change in Water Variable (mg/l)	Odds Ratio[a]	90% Confidence Interval
Model 1			
Hardness	-100	1.20	(1.00, 1.44)
Sodium	15	1.21	(1.05, 1.41)
Model 2			
Alkalinity	-100	1.31	(1.03, 1.65)
Sodium	15	1.28	(1.11, 1.49)
Model 3			
Calcium	-50	1.43	(0.92, 2.22)
Sodium	15	1.21	(1.04, 1.41)
Model 4			
Magnesium	-20	1.36	(1.03, 1.78)
Sodium	15	1.28	(1.06, 1.42)

[a]The values given for the odds ratio are the odds ratio for that
variable given a fixed level of all the other independent variables in the regression. The regression models are given in
Table 7. For the odds ratios presented under cerebrovascular
disease, the two variables presented together are from the
same regression model.

The results for the low sodium (nonsoftened water) group are different from the results for the high sodium (artificially softened water) group, as shown in Table 7. In the high sodium group, there is no apparent difference between cases and controls in either alkalinity or hardness, nor is sodium significantly different. The variables which are statistically significant in this group are metals, with cases having higher levels of both zinc and copper. The results for calcium and magnesium parallel those for alkalinity and hardness. In the low sodium group, both variables are highly statistically significant in their respective regressions, whereas in the high sodium, neither calcium nor magnesium is significant. This may be due to some differences in the softening processes in the two groups rather than sodium levels. In the low sodium group, alkalinity and hardness show greater differences between cases and controls than when cases and controls are not stratified by sodium values.

The nature of the results indicates that the observed association between hardness variables and cardiovascular disease in the low sodium group is not due simply to the use of artificial softening. The association may exist only for those who are not using water softeners. The lack of an observed association in those who do use softeners may be due to negative confounding or the lack of power to detect an association for the small sample size in this group.

Because there were few entrants with high sodium levels, the analysis was confined by cause of death to the group of persons with sodium below 10 mg/l, as shown in Table 8. For both cause-of-death categories, the results are similar to the results observed for the total groups (the combined causes of death). For CAD's, hardness, alkalinity, calcium, and magnesium are statistically significant and are associated with case status. For CBVD deaths, on the other hand, magnesium is statistically significant, while calcium is not (p .32). This is similar to the results obtained when CBVD was analyzed without stratifying by sodium values. For the CBVD deaths, potassium is significantly higher in the cases than in the controls, when the analysis is confined to the low sodium group. This association did not appear when both high and low sodium groups were included. Since the mean potassium level in the CBVD and control groups combined is less than 2 mg/l, the likelihood is small that this level of potassium intake from drinking water is causally related to CBVD risk.

Geographic Location

Ground water varies by geography, and the hardness of well water supplies can be very different within small distances. The range of hardness within a county in this study was generally large. As previously noted, cases and controls were not

Table 7. Regression Models for Softened and Nonsoftened Drinking Water.[a]

Regressions Containing Hardness

Low Sodium Group			High Sodium Group		
Independent Variable	Beta Coeff.	p-Value	Independent Variable	Beta Coeff.	p-Value
Model 1[b]			Model 1[b]		
Age	0.2027	.0001	Age	0.3401	.0001
Hardness	-0.0026	.01	Hardness	-0.0007	.78
Sodium	-0.0391	.39	Sodium	0.0058	.72
Potassium	0.0530	.18	Potassium	0.0403	.27
pH	0.4478	.10	pH	-0.2515	.80
Copper	0.0723	.46	Copper	0.8601	.03
Zinc	0.0683	.56	Zinc	2.5200	.05

Regressions Containing Alkalinity

Low Sodium Group			High Sodium Group		
Independent Variable	Beta Coeff.	p-Value	Independent Variable	Beta Coeff.	p-Value
Model 2[b]			Model 2[b]		
Age	0.2033	.0001	Age	0.3399	.0001
Alkalinity	-0.0036	.0015	Alkalinity	-0.0001	.99
Sodium	-0.0372	.42	Sodium	0.0068	.67
Potassium	0.0568	.15	Potassium	0.0429	.26
pH	0.5379	.05	pH	-0.1654	.86
Copper	-0.0755	.43	Copper	0.8922	.02
Zinc	-0.0677	.56	Zinc	2.4900	.05

Table 7. (continued)

Regressions Containing Calcium

	Low Sodium Group			High Sodium Group	
Independent Variable	Beta Coeff.	p-Value	Independent Variable	Beta Coeff.	p-Value
	Model 3b				Model 3b
Age	0.2023	.0001	Age	0.3400	.0001
Calcium	-0.0115	.02	Calcium	-0.0010	.92
Sodium	-0.0440	.33	Sodium	0.0066	.68
Potassium	0.0536	.17	Potassium	-0.0421	.26
pH	0.3975	.13	pH	0.1989	.84
Copper	-0.0668	.49	Copper	0.8820	.03
Zinc	0.0677	.56	Zinc	2.5060	.05

Regressions Containing Magnesium

	Low Sodium Group			High Sodium Group	
Independent Variable	Beta Coeff.	p-Value	Independent Variable	Beta Coeff.	p-Value
	Model 4b				Model 4b
Age	0.2028	.0001	Age	0.3401	.0001
Magnesium	-0.0225	.01	Magnesium	-0.0089	.67
Sodium	-0.0381	.41	Sodium	0.0049	.76
Potassium	0.0516	.19	Potassium	0.0399	.28
pH	0.4595	.09	pH	-0.2651	.78
Copper	-0.0700	.47	Copper	0.8486	.03
Zinc	0.0674	.56	Zinc	2.5289	.05

[a]Softened Water = High Sodium Group (greater than 40 mg/l sodium); 46 cases, 101 controls.
Nonsoftened Water = Low Sodium Group (less than 10 mg/l sodium); 333 cases, 585 controls.
[b]Cr, Fl, NO_3, Cd, Pb, Fe, Mn included in model, not statistically significant at $p \leq .01$.

Table 8.   Logistic Regression Models by Cause of Death Category
           Including Only Water Supplies with Sodium Less Than
           10 mg/1.

Coronary Artery Disease Deaths
(Cases, N = 268; Controls, N = 584)

Independent Variable	Beta Coeff.	p-Value	Independent Variable	Beta Coeff.	p-Value
	Model 1[a]			Model 2[a]	
Age	0.1989	.0001	Age	0.1996	.0001
Hardness	-0.0023	.03	Alkalinity	-0.0031	.01
Sodium	-0.0280	.55	Sodium	0.0270	.57
Potassium	0.0303	.52	Potassium	0.0333	.48

Coronary Artery Disease Deaths

Independent Variable	Beta Coeff.	p-Value	Independent Variable	Beta Coeff.	p-Value
	Model 3[a]			Model 4[a]	
Age	0.1935	.0001	Age	0.1939	.0001
Calcium	-0.0097	.05	Magnesium	-0.0197	.02
Sodium	-0.0329	.49	Sodium	0.0271	.57
Potassium	0.0303	.52	Potassium	0.0285	.55

Cerebrovascular Disease Deaths
(Cases, N = 65; Controls, N = 584)

Independent Variable	Beta Coeff.	p-Value	Independent Variable	Beta Coeff.	p-Value
	Model 1[a]			Model 2[a]	
Age	0.2719	.0001	Age	0.2709	.0001
Hardness	-0.0033	.15	Alkalinity	-0.0054	.03
Sodium	-0.1370	.12	Sodium	0.1426	.11
Potassium	0.1609	.01	Potassium	0.1666	.01

Table 8. (continued)

Cerebrovascular Disease Deaths

Independent Variable	Beta Coeff.	p-Value	Independent Variable	Beta Coeff.	p-Value
	Model 3[a]			Model 4[a]	
Age	0.2711	.0001	Age	0.2695	.0001
Calcium	-0.0099	.32	Magnesium	-0.0358	.06
Sodium	-0.1482	.09	Sodium	0.1293	.15
Potassium	0.1616	.01	Potassium	0.1585	.01

[a]Cr, Fl, NO_3, Cd, Pb, Fe, Cu, Zn, Mn, pH included in model; not statistically significant.

matched for geographic location because of the concern that this would overmatch for water characteristics.

In order to determine whether geographic location was related to case status, logistic regressions were carried out for each of the three areas (South, Central, and North). The statistical test for differences in logistic models among the three areas is not significant for any of the models using each of the four water-hardness variables. However, the relationship of case status to total hardness and its components, calcium and magnesium, is stronger (as measured by the beta coefficient in the logistic model) and is closer to statistical significance in the Central and North areas than in the South. The South contains the majority of cases and controls, and also has the hardest water.

Another means of analyzing the effect of geographic region is to include area as a variable in each logistic regression model. This was done for each of the models, using each of the four water-hardness variables, with the following results:

o   Area was found not to be a statistically significant variable in any of the models.
o   The p-value for hardness in Model 1 with area added was $p < .02$ compared to a p-value for hardness of < 0.04 when area was not included and the beta coefficient (slope) for hardness in the regression was not changed materially.

o   The addition of area as an independent variable did not
    reduce the relationship of case status with alkalinity
    ($p<.008$) or magnesium ($p<.03$).
o   The p-value for calcium in Model 3 with area added was
    $p<.12$; calcium was not significantly related to case
    status, either with or without the inclusion of area.
o   The maximum likelihood ratio test for the difference be-
    tween models with and without the term "Area" is not
    significant for all four models. Despite the seeming
    differences between the South and the other two areas,
    the nonsignificance of the maximum likelihood ratio test
    for differences among the areas for a given regression
    model indicates that geography does not explain the dif-
    ference in water variables between cases and controls.

Diet

If a higher proportion of controls than cases are dairy
farmers, then it might be expected that intake of dairy products (and hence calcium) could be markedly different in controls than in cases. There was no evidence of any relationship
between level of calcium in drinking water and calcium intake
from food, within either the case or control group. Thus, food
calcium cannot be a confounder of the observed association between case status and water calcium. However, there is ample
reason to believe that food calcium might be a modifier of the
effect of water calcium. Indeed, there are substantial differences between cases and controls in calcium intake from
food, even within age groups. These differences in food intake
may be caused by differences in activity of the farm, or more
likely by changes in dietary habits of cases who had histories
of cardiovascular disease or cerebrovascular disease, or diagnoses of hypertension. It is not expected that drinking water
characteristics of cases would have changed in response to such
a diagnosis or history.

In order to test the possibility that food calcium intake
was an effect-modifier of the association, a logistic regression including this variable was fit. The analysis was limited
to the low sodium group because of potential confounding bias
in the group with home water softeners (Na greater than 40
mg/1). The independent variables in the model included age,
calcium as the hardness variable, and an estimate of total
daily calcium intake from dairy products, as shown in Table 9.
Estimated intake of calcium from dairy products is considerably
lower in the case group, but this does not alter the significant relationship previously observed for water calcium.

Table 9.   Logistic Regression Model Containing Estimated Intake
of Calcium from Dairy Products.

Independent Variable	Beta Coeff.	p-Value
Age	0.1958	.0001
Dairy Calcium[a]	-0.0010	.0003
Water Calcium	-0.0102	.03
Water Sodium	-0.0339	.45
Water Potassium	0.0454	.24
Water Chromium	86.28	.52
Water Fluoride	-0.1480	.80
Water Nitrate	-0.0046	.85
Water Cadmium	169.6	.15
Water Lead	-2.536	.76
Water Iron	-0.0837	.37
Water Copper	-0.0463	.63
Water Zinc	0.0300	.80
Water Manganese	1.168	.24

[a]Measured in average intake per day in milligrams.

Other Potential Confounders

Information about cigarette smoking was obtained from telephone interviews or the mailed questionnaires. Smoking (measured in average packs per day) was not related to any water
parameter, and inclusion of smoking in the logistic regressions
had no effect on the estimates for water variables. Therefore,
smoking is not a confounder of the observed association with
any water variable. However, smoking was a strong risk factor
for case status (estimated odds ratio of 2.3 for smoking one
pack per day for the previous five years).
Measurement of exercise level and stress, along with other
potential cardiovascular disease, was not possible in the present study. However, there is no reason to believe these risk
factors varied with drinking water characteristics over the
statewide geographic region from which cases and controls were
drawn.

DISCUSSION

The consistent difference between the case and the control
group in this cohort of Wisconsin farmers is that cases have

lower hardness, alkalinity, and probably lower calcium in their daily drinking water. These associations are strengthened when only participants with low (less than 10 mg/l) sodium content are considered. It is felt that these participants do not artificially soften water with home ion exchange units. There is no apparent relationship between hardness variables and case status among participants with high (more than 40 mg/l) sodium levels. It appears that these participants artificially soften water, which may confound the association or offer too few participants to detect an association. Based on the measurement of sodium level, CAD cases have fewer home water softeners than controls, while CBVD cases have more home water softeners than controls.

CBVD cases have higher sodium levels than do either controls or CAD cases. This raises the possibility that the softer drinking water in that group is due primarily to a higher level of softener use. When only persons with low sodium levels are considered, lower magnesium levels were found in CBVD cases than in controls, but not lower calcium levels.

For several of the individual metals, cases have significantly different levels than controls in certain subgroups, but the relationship is not consistent across groups and some factor in the softening process might be responsible for this finding. The higher potassium levels found in CBVD cases who do not soften the drinking water is also without obvious explanation. The likelihood that actual intake of potassium from drinking is directly related to risk is small, because the levels of potassium in drinking water are minimal compared to intake from dietary sources [16].

In general, the relationships found in the CAD group are more straightforward and easily interpreted than those for CBVD. CAD cases without artificial softeners have lower levels of calcium and magnesium in their drinking water, and lower hardness and alkalinity. There are no other readily apparent differences in other water constituents measured. While the possibility exists that unidentified confounding explains these relationships, such confounding is not readily apparent. For CBVD cases, a larger case group is really needed for detailed analysis and interpretation.

Traditionally, the observation of higher cardiovascular deaths in soft water areas has led to suggestions that either hard water contains something beneficial or soft water contains something deleterious, or possibly both. While it is not possible from this single analytic epidemiologic study to draw definitive conclusions, the results tend to lend credence to the idea that the "softer water-higher cardiovascular risk" relationship may indeed be real and that an association exists between CAD and the calcium and magnesium of drinking water content. While the magnitude of this association is apparently not large, the potential exists for the prevention of cardiovascular disease in a large number of people by changing water treatment practices, if this association is causal. An association between cardiovascular disease and metals in drinking water was not ruled out by this study, as the levels of lead and cadmium, the prime suspects, were very low in the study participants' water supplies. This obviated any opportunity to examine high intake of these metals.

ACKNOWLEDGMENT

The research described in this chapter was sponsored by the Environmental Protection Agency, Health Effects Research Laboratory, Cincinnati, Ohio, under Interagency Agreement No. 40-1063-80 with Martin Marietta Energy Systems, Inc., under Contract No. DE-AC05-84OR21400 with the U.S. Department of Energy.

REFERENCES

1. Kobayashi, J. "On Geographic Relationship between the Chemical Nature of River Water and Death Rate from Apoplexy," Berichte des Ohara Institut für Landwirtschaftliche Biologie 11:12-21 (1957).

2. Folsom, A. R., and Prineas, R. J. "Drinking Water Composition and Blood Pressure: A Review of the Epidemiology," Amer. Jour. Epidemiol. 115:818-832 (1982).

3. Masironi, R., and A. G. Shaper. "Epidemiological Studies of Health Effects of Water from Different Sources," Ann. Rev. Nutr. 1:375-400 (1981).

4. Comstock, G. W. "The Epidemiologic Perspective: Water Hardness and Cardiovascular Disease," Jour. Environ. Path. Toxicol. 4:9-25 (1980).

5. National Research Council. Drinking Water and Health, Vol. 3 (Washington, D.C.: National Academy Press, 1980), pp. 21-24.

6. Comstock, G. W. "Water Hardness and Cardiovascular Diseases," Amer. Jour. Epidemiol. 110:375-400 (1979).

7. Sharrett, A. R. "The Role of Chemical Constituents of Drinking Water in Cardiovascular Diseases," Amer. Jour. Epidemiol. 110:401-419 (1979).

8. National Research Council. Geochemistry of Water in Relation to Cardiovascular Disease (Washington, D.C.: National Academy of Sciences, 1979).

9. Neri, C. C., D. Hewitt, and G. B. Schreiber. "Can Epidem-
    iology Elucidate the Water Story?" Amer. Jour. Epidemiol.
    99:75-88 (1974).

10. Neri, C. C., D. Hewitt, G. B. Schreiber, T. W. Anderson,
    J. S. Mandel, and A. Zdrojewsky. "Health Aspects of Hard
    and Soft Waters," Jour. Amer. Water Works Assoc. 67:403-
    408 (1975).

11. Comstock, G. W. "Fatal Arteriosclerotic Heart Disease,
    Water at Home, and Socio-economic Characteristics," Amer.
    Jour. Epidemiol. 94:1-10 (1971).

12. Shaper, A. G., R. F. Packham, and S. J. Pocock. "The
    British Regional Heart Study: Cardiovascular Mortality
    and Water Quality," Jour. Environ. Path. Toxicol. 4:89-111
    (1980).

13. Sharett, A. R., M. M. Morin, R. R. Fabsitz, and K. R.
    Bailey. "Water Hardness and Cardiovascular Mortality," in
    Drinking Water and Human Health, J. A. Bell and T. C.
    Doege, Eds. (Chicago, IL: American Medical Association,
    1984).

14. "Census of Agriculture, 1974." Bureau of the Census (U.S.
    Government Printing Office, 1976).

15. Supplemental Library Users Guide, 1980 Edition. (Gary,
    NC: SAS Institute, 1980).

16. Safe Drinking Water Committee. "Inorganic Solutes," in
    Drinking Water and Health (Washington, DC: National Aca-
    demy of Sciences, 1977).

CHAPTER 18

EMPIRICAL EXPOSURE MEASURES IN RETROSPECTIVE
EPIDEMIOLOGIC STUDIES

Charles E. Lawrence and Philip R. Taylor

INTRODUCTION

The accurate assessment of exposure in epidemiologic studies can have an important impact on the results of these studies. Of course, if exposure assessment in the diseased and
nondiseased study subjects differs, a bias is induced. Equally
inaccurate assessment of exposure in the two groups results in
equal misclassification. As pointed out by Bross [1], this results in a loss of power but does not affect the size of the
statistical test of association between disease and exposure.
As a consequence, important disease exposure associations can
be overlooked.

Latency periods of 20 years or more, which are common in
chronic diseases, pose substantial difficulties for accurate
estimation of exposure in epidemiologic studies. Since actual
exposure measures are rarely available over a 20-year period,
most epidemiologic studies are forced to use surrogate measures
of exposure, usually broad classifications that are derived deductively. One such measure is used in studies relating chloroform in drinking water to cancer. Individuals are typically
classified as "exposed" if their home water supply at the time
of diagnosis (for incident-based studies) or death (for
mortality-based studies) was a chlorinated surface water
source, or "unexposed" if the source at that time was either an
unchlorinated surface source or a groundwater source.

Here we propose an alternative approach for retrospective
epidemiologic studies which uses empirically based estimates of
exposure. As examples, we describe two studies: a) the use of

an empirical model for estimating cumulative exposure to chloroform from drinking water, which was applied to a case-control study of colorectal cancer and drinking water; and b) an empirical estimate of serum polychlorinated biphenyl (PCB) concentrations, which was used in a cohort study of the relation of PCB to pregnancy outcomes.

TRIHALOMETHANES IN DRINKING WATER AND COLORECTAL CANCER

In 1980 we undertook a case-control study of the relation of trihalomethanes (THM) in drinking water to colorectal cancer. The human carcinogenic potential of THM became a matter of concern when chloroform was identified as an animal carcinogen and when Rook demonstrated that THM are produced by the chlorination of drinking water [2-6]. Previous epidemiologic studies had reported an association, based primarily on ecologic analyses, between consumption of chlorinated drinking water and the prevalence of cancer at various sites, including the colorectum [7-11]. We were particularly concerned with two aspects of these previous reports: exposure was usually crudely categorized into only two groups, chlorinated versus nonchlorinated drinking water; and controls for confounding variables seemed inadequate.

A potential confounding factor of special concern was population density, since there is a well-known urban-rural gradient in both colon cancer and water chlorination. To address this problem, we selected all of our cases and controls from the New York State Teachers Retirement System, which includes all public school teachers in the state outside of New York City. The resulting study group was highly homogeneous in occupation and socioeconomic status, yet geographically dispersed in a manner similar to the total population of the state. Thus, their range of exposure to chloroform in drinking water could be presumed to parallel that of the general population of the state.

The ideal estimate of exposure for this study would have been determined from actual measurements of drinking water across the state over the biologically relevant time period. Unfortunately, because of the estimated 20-year latency period associated with colorectal cancer, such data were not available. Indeed, over most of this interval, the presence of chloroform in drinking water had not yet been reported, nor had the analytic methods to measure chloroform in such low concentrations been developed.

In 1978 the New York State Department of Health completed a survey of THM in public drinking water sources in the state including 174 water supply systems [12]. However, many of the water supply systems used by study subjects were not covered by the survey. More importantly, water treatment conditions affect the formation of chloroform, and these conditions had changed substantially in the years prior to the survey. It was exposure to chloroform in these prior years that we were most interested in.

Our solution was to develop a multivariate regression model, based on the water-treatment survey data, which would allow us to derive an empirical measure of chloroform exposure. The details of the regression analysis are presented elsewhere [13]. The final regression equation was:

$$\log_e Y = 2.59 + 0.35 \log_e X_1 + 0.41 \log_e X_2 + 0.11 \log_e X_3$$

where Y = chloroform concentration ($\mu g/l$),

X_1 = prechlorine plus postchlorine dose (pounds/million gallons)

X_2 = effluent chlorine residual plus 0.25 ($\mu g/l$),

and

X_3 = source type (1 = lake, 2 = stream, 3 = river, and 4 = reservoir).

The R^2 for this model was 0.54 based on 164 observations. Transformation to log scale for the dependent variable was required to meet the assumption of normality and homoscedasticity of the residuals. This transformation was in agreement with the finding that chloroform concentrations are lognormally distributed.

This regression equation provided a means to estimate the expected chloroform concentration for each of the study subject's water supplies for specific years of interest. From existing records of the Teachers Retirement System and local schools, we constructed residential and work location histories for all of our study subjects (395 cases and 395 controls). We also obtained records from water treatment plants serving these locations for the previous 20 years. Assuming an average daily water consumption of 2 liters (1 liter at work and 1 liter at home), we used the expected concentrations to calcuate an expected cumulative lifetime dose. No significant difference in estimated chloroform exposure was found between cases and controls, as shown in Figure 1.

PREGNANCY OUTCOMES IN WOMEN OCCUPATIONALLY EXPOSED TO PCB

In a cohort study of the relation of PCB exposure to reproductive outcomes, we were again confronted with the problem of historical assessment of exposure levels [14]. In this case, the latency period was no more than 9 months, but the births to the women in the cohort occurred over a 34-year period from 1949 to 1983. The actual serum PCB concentrations in these mothers during each of their pregnancies were, of course, unknown.

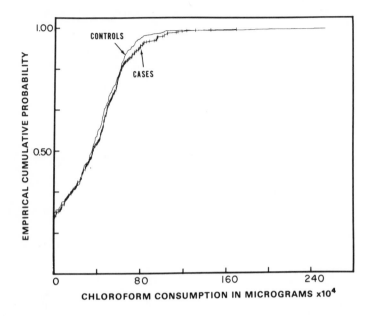

Figure 1.   Empirical distribution function of cumulative lifetime dose of chloroform (see Reference 7).

For each employee, however, we were able to obtain from her
employer a complete work history, specifying the job she held
during each month she worked at the plant.   Industrial hygiene
data on PCB and job-process information enabled us to classify
each job in the plant as involving either indirect (at the facility but not in the production areas) or direct exposure.   Direct exposure jobs were subcategorized as low (air contact only) or as medium, variable, or high (air contact plus increasing degrees of dermal contact).
The resulting data set provided us with a wide range of PCB
exposure surrogates, including direct versus indirect, highest
level ever exposed, total months at any direct exposure, and
total months employed.   The challenge was to choose a surrogate
measure that best approximated the ideal (but unavailable)
measure.
Fortunately, sera from 152 employees (118 men and 34 women)
of the plant had been analyzed for PCB concentration in 1976 as
part of an evaluation by the company of general health and PCB
exposure.   From these data we could empirically estimate the
relation between the employment history variables and serum
high-homolog PCB concentrations.   Regression analysis led to
the model:

$$\log_e Y = -0.850 + 0.259 \; X_1 + 0.026 \; X_2$$
$$+ \; 0.069 \; \log_e X_3 + 0.673 \; \log_e X_4$$

where Y = serum Aroclor™ 1254 (parts per billion),

           X_1 = sex (0 = female, 1 = male),

           X_2 = age (years),

           X_3 = weighted cumulative number of months worked from 11.5 to 21.5 years before the blood sample was taken,

and

           X_4 = weighted cumulative number of months worked from 0 to 11.5 years before the blood sample was taken.

The R^2 for this model was 0.64. Serum PCB concentrations were lognormally distributed. The dependent variable was log-transformed in the model to meet the assumptions of normality and homoscedasticity of the residuals.

The weights for variables X_3 and X_4 were derived by analysis of the data and pertain to the effects of these levels of exposure on serum PCB concentrations for a given job. Several weighting schemes were tested, but no significant advantage was achieved over the simple scheme: 0 = not employed, 1 = indirect, 2 = low, 3 = medium, 4 = variable, and 5 = high.

This model allowed us to estimate the expected serum high-homolog PCB concentration for each woman during each of her pregnancies. No association was found between these concentrations and birthweight or gestational age.

DISCUSSION

The credibility of an empirically derived surrogate, such as those presented here, should be judged in comparison to that of alternative, deductively derived surrogates for the specific study. If the existing literature provides a sound conceptual framework for the use of a deductively derived surrogate, an empirically derived surrogate may be either unnecessary or inferior. The absence of such a conceptual framework may make an empirically derived surrogate preferable. Its credibility rests on how closely the exposure sample used to derive the model resembles the population of the subsequent epidemiologic study.

The primary limitations in the use of empirical measures of exposure stem from limitations in the representativeness of the exposure sample. In the THM study, the exposure sample was limited by the limited number of seasonal water samples available, the need to assume a volume of water consumed by each study subject, and by the time differences between the collection of the water samples and the time of exposure of the study subjects. In the PCB study, the exposure sample was limited by the low proportion of women in the exposure sample and the unavailability of serum samples in earlier eras when engineering control of exposure was more limited.

The final choice between these two types of exposure surrogate for a given study is not reached statistically, it is a judgment of the relative credibilities of the surrogates. For example, our choice in the PCB study was based on our judgment that in spite of the limitation of the exposure sample, the model of measured serum concentrations would provide a more credible basis for assessing exposure than any deductive inference from the existing literature on metabolism of PCB in humans.

A further advantage of an empirically derived surrogate is that the exposure data themselves provide an objective basis for judging a proposed model. This process makes any shortcoming of the model explicit. For example, the uncertainty in the predicted exposure can be described by parameters of the model, such as the R^2 or confidence intervals. In contrast, for deductively derived indices, no objective or quantitative basis for judging the proposed index generally exists.

DISCLAIMER

The work described in this chapter was not funded by EPA and no official endorsement should be inferred.

REFERENCES

1. Bross, I. "Misclassification in 2 x 2 Tables," *Biometrics* 10:478-486 (1954).

2. Eschenbremer, A. B., and E. Miller. "Induction of Hepatomas in Mice by Repeated Oral Administration of Chloroform with Observations on Sex Differences," *J. Natl. Cancer Inst.* 4:251-255 (1945).

3. Rudali, B. "A Propos de l'Activitie Oncogene de Quelques Hydrocarbures Halogenes Utilises Entheropentique," *UICC Monogr. Series* 7:138 (1967).

4. Roe, F. J., F. L. Carter, and B. C. Mitchley. "Tests of Miscellaneous Substances for Carcinogenesis. Test of Chloroform and 8-Hydroxyquinoline for Carcinogenicity Using Newborn Mice," *Br. Emp. Cancer Campgn. Res. Annu. Rep.* 56:13 (1978).

5. Page, N. P., and U. Saffiotti. "Report on Carcinogenesis Bioassay of Chloroform," National Cancer Institute (1976).

6. Rook, J. J. "Formation of Haloforms during Chlorination of Natural Waters," J. Soc. Water Treatment Exam. 23:234-243 (1974).

7. Harris, R. S. The Implications of Cancer-Causing Substances in Mississippi River Water (Washington, DC: Environmental Defense Fund, 1974).

8. Kuzma, R. J., C. M. Kuzma, and C. R. Bunchar. "Ohio Drinking Water Source and Cancer Rates," Am. J. Public Health 67:725-729 (1977).

9. Kruse, C. W. "Chlorination of Public Water Supplies and Cancer: Washington County, Maryland, Experience," Preliminary Report, EPA Grant No. R805198-01-0. (Cincinnati, OH: Environmental Protection Agency Health Effects Research Laboratory, 1977).

10. Cantor, K. P., R. Hoover, T. J. Mason, and L. J. McCabe. "Associations of Cancer Mortality with Halomethanes in Drinking Water," JNCI 61:979-985 (1978).

11. Hogan, M. D., P. Y. Chi, D. G. Hoel, and T. J. Mitchell. "Association Between Chloroform Levels in Finished Drinking Water Supplies and Various Site-specific Cancer Mortality Rates," J. Environ. Pathol. Toxicol. 2:873-887 (1979).

12. Schreiber, J. S. "The Occurrence of Trihalomethanes in Public Water Supply Systems in New York State," J. Am. Water Works Assoc. 73:154-159 (1981).

13. Lawrence, C. E., P. R. Taylor, B. J. Trock, and A. A. Reilly. "Trihalomethanes in Drinking Water and Human Colorectal Cancer," J. Natl. Cancer Inst. 72:563-568 (1984).

14. Taylor, P. R., J. M. Stelma, and C. E. Lawrence. "The Relation of Polychlorinated Biphenyls to Birthweight and Gestational Age in the Offspring of Occupationally Exposed Mothers," Report to the National Institute for Occupational Safety and Health, (1984).

CHAPTER 19

EVALUATION OF LEAD EXPOSURES IN THE ENVIRONMENT AND
THEIR CONTRIBUTION TO BLOOD LEAD LEVELS IN CHILDREN

Daniel Greathouse

INTRODUCTION

This paper presents the results of an epidemiologic study
designed to assess the contribution of lead in drinking water
to lead exposure in infants. Since infants are surrounded by
potential sources of lead intake throughout their lives, the
drinking water contribution cannot be considered as a one-time
occurrence or in isolation from other potential sources such as
air, dust, and food. For these reasons, a longitudinal study
was conducted which involved repeated assessments of lead levels in the blood and household environments of pregnant women
and their infants. Changes in infant blood lead levels during
the first two years of life are related to the average levels
of lead observed in each of the measured sources.

MATERIALS AND METHODS

Pregnant women living in the vicinity of Columbus, Ohio,
and Boston and New Bedford, Massachusetts, who received pre-
natal care during 1978-79 from selected clinics and physicians
and met prespecified criteria (concerning age, length of pregnancy, health, and willingness to make a long-term commitment
to participation) were invited to participate. Repeated blood
samples were collected from most of the mothers during pregnancy and their infants from the time of birth until two years
of age (as shown in Table 1). At least three blood samples

Table 1. Distribution of Number of Lead Measurements Per Infant During the First Two Years of Age.

Sample Type	Percent of Subjects with Less than or Equal to the Specified Number of Lead Measurements													
	0	1	2	3	4	5	6	7	8	9	10	11	12	13
Blood														
Columbus (230[a])	-	-	-	3	9	12	20	24	30	42	65	90	99	100
Boston (199)	-	-	-	16	27	43	55	69	79	88	97	98	100	100
New Bedford (100)	-	-	-	3	4	7	15	27	51	78	96	100	100	100
Tap Water														
Columbus (232)	1	9	23	40	62	81	95	100	100	100	100	100	100	100
Boston (193)	1	6	21	34	55	71	84	92	100	100	100	100	100	100
New Bedford (109)	0	3	5	9	2	23	44	64	97	98	100	100	100	100
Household Air														
Columbus (232)	1	9	25	41	66	84	94	100	100	100	100	100	100	100
Boston (193)	1	8	21	37	59	73	85	93	100	100	100	100	100	100
New Bedford (109)	0	3	6	10	17	28	43	72	97	100	100	100	100	100
Household Dust														
Columbus (232)	0	8	27	41	63	85	95	100	100	100	100	100	100	100
Boston (193)	1	6	21	35	54	70	85	92	100	100	100	100	100	100
New Bedford (109)	0	3	6	8	13	19	41	64	97	100	100	100	100	100

[a]Total number of participants who donated subject samples from each community.

were drawn from 530 infants (233 from Columbus, 188 from Boston, and 109 from New Bedford) during the first two years of life. Columbus, Boston, and New Bedford were selected to represent urban areas with a gradient of exposure levels to waterborne lead. Multiple samples of household drinking water, dust, and air particulate were collected from the participant residences during the 2-3 year observation period (Table 1); many of these samples were collected near, but not necessarily at the same time, as the blood samples. All samples were analyzed for lead content. Levels of lead in the diets of the women and their infants were estimated from 24 dietary recalls that were coded for lead content using published information from the U.S. Food and Drug Administration.

All blood samples from the pregnant women consisted of 4 ml of venous blood drawn with a 5 cc lead-free disposable plastic syringe. Infant blood samples (minimum of 0.2 cc of whole blood) were collected by finger or heel stick until age 6-12 months, and by venous sampling for subsequent draws. Collection containers were routinely sampled and tested for lead contamination. The samples from Boston and New Bedford were packed in dry ice and mailed within one week of collection to Columbus for analysis [1]. Multiple water samples (7 in Columbus and 9 in other cities) were collected from each residence at each collection time to represent different collection locations (kitchen and bathroom), length of residence time in plumbing pipes (overnight sample, grab sample during the day, running sample after 5 minutes), and the effects of heating water versus using cold water. These samples were collected in 30 ml polyethylene containers, preserved with 1 ml of nitric acid (sufficient to reduce pH to 2.0-3.5), and sent for analysis within 2-3 weeks of collection. Air particulate samples were collected on membrane filters (0.8 μM pore-size, Teflon^{TM} coated) with a portable pump placed for 24 hours in the bedroom or play area of the infant; the house dust samples were collected with the same portable pump from a 50 cm x 50 cm area in a high traffic area of the residence [2]. All blood and environmental samples were collected in containers supplied by a laboratory at Children's Hospital in Columbus, Ohio, and analyzed by atomic absorption spectrophotometry (Instrumentation Laboratories, Model 251) by the same laboratory. The flameless method was used for the blood samples, due to the small sample size requirements; lead values (reported in μg/dl) are the mean of at least two analyses on a diluted sample (one in 10 dilution using Brig 35, 0.1% solution). Throughout this study the Columbus laboratory remained consistently within the top third of the laboratories participating in the Center for Disease Control Blood Lead Proficiency Testing Program. The water, air particulate, and dust samples were analyzed by the flame method and reported respectively in units μg/l, μg/m^3, and μg/g. Working standards (prepared from stock supplied by Fisher Scientific Company, Pittsburgh, PA, at lead concentrations of 1, 50, 100, 500, and 1000 ppb) were used to generate a new standard curve each day and were rerun after every 12 unknowns to check for machine drift [1].

RESULTS

In each community, blood lead levels tended to be low at birth (median levels of 8-9 µg/dl), and increased during the first two years of life. The communities differed, however, in terms of the amount of change during the two years and the distribution of blood lead levels about the observed medians. In Boston, the median increased to approximately 21 µg/dl at 2 years, versus 14 µg/dl in both Columbus and New Bedford. The distribution of blood lead levels in Boston appears to be more skewed toward higher values than in the other two communities, indicating a greater frequency of high levels in Boston than in the other two communities. As would be expected, these trends are generalizations that do not adequately describe fluctuations in the median blood lead levels over time nor the individual measurements about these medians.

Recognizing these general trends in infant blood lead levels and the individual differences about these trends, the question is how to assess the relative contribution of different environmental sources of lead to these levels. Environmental lead exposure is not a one-time event from one environmental source, but includes several potential contributors throughout an infant's life. Starting with in utero exposure, potential environmental lead sources that may contribute throughout life include drinking water, air, dust, food, and paint. Mothers' blood lead tends to be low and relatively constant during pregnancy, near the levels observed in the infant at the time of birth, hence it is unlikely that mothers are significant contributors to the pattern of increasing blood leads observed in their infants.

As explained earlier, the primary objective of this study was to assess the contribution of lead in drinking water to blood lead levels, and the three communities were selected to represent a gradient of water lead exposures. In general, three levels of water lead exposure are represented by Columbus with the lowest levels, Boston with intermediate levels, and New Bedford with the highest levels. However, there is considerable variation in lead content among different types of water samples and individual samples within each type, particularly in New Bedford, which had corrosive drinking water and lead service lines at the time of this study. Given this variation in lead content, it seems clear that the contribution of water lead to blood lead will depend on the pattern of water usage and residence locations during infancy. For example, if a family moves from one residence with a certain length of lead service line to another with a different length of lead line or without a lead service line, the contribution of water lead will change. Also, the pattern of water usage may change during infancy. Infant formula may be prepared with hot water, while water consumed directly may be drawn from the cold water faucet, or the length of time the water is allowed to run prior to filling a drinking container may vary from one occasion to another. All these factors influence the levels of water lead intake. Another consideration likely to be important is that the quantity of water consumed per body weight of the infant will probably decline as the infant starts eating more solid food.

Levels of lead in household air and house dust also vary among the three communities, but not in the same gradient as levels in drinking water. New Bedford, with the highest levels of water lead levels, also has high levels of dust lead levels but low levels of airborne lead. On the other hand, Columbus, with the lowest levels of waterborne lead, also has low levels of dust lead but higher levels of airborne lead. As would be expected, these trends are general and there is considerable variation in individual levels observed in the communities.

A further complication to assessing relationships between environmental levels of lead and blood lead is the fact that blood lead represents an accumulation of lead intakes over a period of time, not just the level of intake on one day. Hence there is a need to develop time weighted estimates of integrated exposure levels for each potential environmental source (i.e., a time-weighted average of the quantity of lead in drinking water, air, dust, etc., consumed over the time period represented by the measured blood lead levels), a task fraught with numerous difficulties and uncertainties.

The approach used to explore the relationships between environmental lead sources and blood lead levels is to relate individual changes in blood lead during the first 2 years of life to the mean levels of lead found in the environmental sources. In other words, the blood lead measurements for each individual were summarized by a measure of change (slope) which was regressed against the mean levels of lead found in each respective household environment. These slopes, or changes, in blood lead will likely be due to increased exposure resulting from changes in activities and/or food and water consumption patterns with increasing age, and may also be due in part to lead accumulation over time. Due to the limited number of blood lead measurements for each infant (1-13 samples per infant), the ordinary least square estimates of the slopes would not be very precise, i.e., the associated variances would be large.

As an alternative, an empirical Bayes approach was employed using the information from all infants to improve the slope estimates (i.e., reduce variances) [3, 4]. The underlying assumption of this approach is that each observed individual slope "b_i" is a sample of size 1 from a population "β_i" and that the population of slopes has some underlying distribution (usually normal). Empirical Bayes slopes are weighted averages of the least squares slopes and the overall mean of the population of population slopes with weights proportional to the variances of the individual least squares slope estimates. Hence, empirical Bayes slopes corresponding to least squares slopes with large variances will be weighted heavily towards the overall population mean, and those corresponding to least squares slopes with small variances will be weighted more heavily towards the least squares slopes.

Slope estimation was restricted to infants with at least 3 blood measurements and complete data for all lead sources considered (i.e., at least one measurement for each source). Each community was treated as a separate population, that is, a separate empirical Bayes analysis was performed for each. Weighted least squares was used to fit a separate linear model for each community relating individual least squares slope estimates to the selected independent variables (mean levels of potential lead exposures, a baseline level of blood lead, and a summary time measure), which were assessed for each individual, as shown in Table 2. Weights were appropriately chosen in order to produce empirical Bayes estimates.

These overall models explain approximately 21-36% of the variation in the individual slopes for each community. As would be expected, there is considerable variation in the individual coefficient estimates among the three communities. Note, however, the consistency in signs among the three communities. The only exceptions are for three coefficients that are not statistically significant from zero ($p > .05$) and hence may be due to random variation and/or collinearity among the independent variables; the possibility of collinearity was not formally tested. The signs of all the coefficients which are statistically significant ($p < .05$) are in the expected direction, except for lead in food/body weight.

Assessing the relative contribution of the independent variables to changes in blood lead must be regarded as only exploratory, due to correlations among the variables, different units of measurements, and varying degrees of variation. For example, differentiating between contributions of household air and dust will be very tenuous since they are significantly ($p \leq 0.06$) correlated. Note also that comparisons of coefficient estimates and relative contributions of independent variables among communities are very tenuous due to differences in levels of the independent variables and correlational structures among the communities. For example, the levels of lead in dust for Columbus are roughly one-third to one-half the levels in the other communities, and the correlations with lead in household air are 0.17, 0.17, and 0.40 respectively, for Columbus, Boston, and New Bedford. Notwithstanding these precautions, there is a need for information concerning the relative contributions of possible environmental lead exposures to blood lead.

The approach used in this analysis is to evaluate each variable in terms of the estimated change in blood lead (from the predicted mean level for two years of age in each community) that would result from a one-standard-deviation change in the variable, while holding other variables constant, as shown in Table 3. For example, if the level of lead in water is increased by one standard deviation (0.614385 in logs or 3.81871 in untransformed units) in Columbus, the predicted mean blood at two years of age (18.79 µg/dl) will increase by 0.29 µg/dl to 19.08 µg/dl. Likewise a one-standard-deviation change in household air lead in Columbus will increase the estimated mean blood lead level at two years by 1.77 µg/dl. Table 4 shows example calculations. From these estimated changes it appears

Table 2. Empirical Bayes Estimation of Community Level Models
Relating Changes in Blood Lead During First Two Years
of Age to Lead Exposures and Covariates.

Variable	Columbus (n = 165)	Boston (n = 128)	New Bedford (n = 69)
Intercept (log(μg/dl)/day))[a]	0.3378[c],[d] (0.0004)	0.3988 (0.0013)	0.0831 (0.4965)
Lead in tapwater (log(μg/l))[b]	0.0035 (0.7042)	-0.0099 (0.3841)	0.0026 (0.7780)
Lead in household air (log(μg/m^3))	0.0769 (0.0319)	0.0227 (0.6655)	0.4243 (0.0002)
Lead in household dust (log(μg/g)	0.0179 (0.1115)	0.0259 (0.0334)	0.0062 (0.6388)
Mother's blood lead (log(μg/dl))	0.0211 (0.2130)	0.0206 (0.3373)	0.0355 (0.2440)
Neonate blood lead (log(μg/dl))	-0.1026 (0.0001)	-0.1508 (0.0001)	-0.1173 (0.0001)
Lead in food/body wt. (log(μg/g/g))	-0.0454 (0.0446)	-0.0192 (0.5041)	-0.0085 (0.8266)
Summary time measure (days)	-0.0004 (0.0062)	-0.0003 (0.0033)	-0.0002 (0.4083)

[a]Units of dependent variable (slope).
[b]Units of independent variables (covariates).
[c]Coefficients/(Probability > 0).   [d]x 10^{-2}.

Overall Model			
F value	7.271	10.499	6.564
Probability > 0	0.0001	0.0001	0.0001
Adjusted R^2	0.2112	0.3436	0.3642
Root mean square error	1.1757	1.0174	0.9500
Mean of dependent variable	0.0009	0.0013	0.0009
Coefficient of variation	130,942	77,516	104,568

Table 3.   Change in Estimated Blood Lead Level (µg/dl) that
            Would Result at the End of Two Years from Increasing
            Each Variable by One Standard Deviation While Holding
            the Other Variables Constant.

Variable	Columbus (n = 165)	Boston (n = 128)	New Bedford (n = 69)
Lead in tapwater	0.29	-1.52	0.39
Lead in household air	1.77	0.70	5.76
Lead in house dust	1.35	4.14	0.68
Mother's blood lead	1.10	1.75	1.42
Neonate blood lead	-4.42	-8.59	-5.48
Lead in food/body wt.	-1.62	-1.14	0.29
Summary time measure	-3.64	-5.23	2.12
Overall mean blood lead at two years	18.79	28.98	17.84

that lead in household air contributes roughly 6 times more to
the increase in blood lead than does water lead in Columbus.
Since, however, the coefficient for water is not statistically
significant from zero (p=.7042) this comparison is probably
very unstable.

DISCUSSION

     Due to the increasing concern over the health implication
of low level lead exposure in infants, the need for longitudinal assessments, such as this study, has increased. Most
previous investigations have been of the cross-sectional type
and have provided little information concerning the trend of
blood lead levels in normal infants with increasing age, or adequately assessed the contributions of various environmental
sources to these apparent trends. Mahaffey et al. [5] found in
their preliminary analysis of the data for infants included in
the Health and Nutrition Examination Survey II (HANES II) study
(a national cross-sectional survey) that mean blood lead levels
for male and female infants increase, respectively, from 11.8
and 14.8 µg/dl at 6 months of age to 19.0 and 18.1 µg/dl for
the age group 1 to 3 years. These levels and the apparent increase in blood lead with age appears consistent with the levels and trend observed in this study. The HANES II survey was
conducted in 1976-1978, and the investigation described in this
chapter in 1977-1980, so that comparability would be expected.

Table 4. Example Computation of Estimated Increase in Blood
Lead at the End of Two Years in Columbus if Log Air
Lead Increases by One Standard Deviation.

---

1. Estimated mean blood lead (PbB) in Columbus at two years of
age:

Let log PbB = a + b(age), a linear model relating
individual blood lead measurements to age (in days).

Let a = mean of intercepts for individual ordinary least
squares models (mean of Line 1, Table 2)

b = weighted mean of slopes (dependent variables of
community level model - Overall Model, Table 2)

age = 730 days.

Then by substitution

$PbB = e^{2.2762 + 0.0009(730)} = 18.79 \ \mu g/dl$.

2. Estimated mean blood lead level if log air lead increases by
one standard deviation:

Let log PbB = a + (b + c s)(age)

where c = coefficient of air term in community level
weighted regression (Table 2)

s = standard deviation of log air lead (0.1604
for Columbus).

Then by substitution

$PbB = e^{2.2762 + (0.0009 + 0.000769 \quad 0.1604)(730)}$

$= 20.56 \ \mu g/dl$.

3. Estimated increase in blood lead if log air lead increases
by one standard deviation:

Increase = 20.56 - 18.79 = 1.77 $\mu g/dl$.

Most, if not all, investigations that have examined the relationship of potential lead exposures and sociodemographic in-
fluences with elevated blood lead have involved some type of
comparison (usually pairwise) of cross-sectional determinations

of blood lead levels, measurements of lead in environmental sources, and survey data [6].

From both the pairwise comparison studies and those that simultaneously considered multiple risk factors in addition to the measured environmental sources, it is apparent that there are multiple factors besides the environmental sources (such as air, dust, drinking water, etc.) that influence levels of blood lead found in children [6, 7, 8, 9]. In an investigation of 377 children living in New Haven, Connecticut, only 10.5% of the variation in blood lead could be explained by measured sources of environmental lead (exterior air lead, house dust, interior and exterior paint, and soil) [7]. Sociodemographic characteristics that have been associated with elevated blood lead levels include low socioeconomic status, disturbed mother-child relationship, frequent moves, single parent families, underemployment, large family size, inadequate parental supervision, and cultural acceptance or encouragement of oral gratification as a means of relieving anxiety [7]. Age, sex, and race have also been identified as important factors influencing levels of observed levels of blood lead [5, 7].

The multiplicity of interrelated contributing factors to blood lead probably at least partially explains the reasons for the relatively low percentage (21-36%) of variation in blood lead slopes explained by the levels of lead in environmental sources. Other probable reasons for the low percent include the lack of quantitative estimates of intake from environmental lead sources, changes in residence during the observation period, and important sociodemographic characteristics that influence both nutrition and behavior which have not been ascertained.

REFERENCES

1. Kranjc, B. B. "Water Intake and Other Environmental Sources of Lead as Related to Body Burden of Lead in Children," MS Thesis, Graduate School of the University of Massachusetts, Amherst, MA (1983).

2. Caffo, A. L., A. H. Lubin, and C. M. Baldeck. "An Inexpensive Pump for Routine Environmental Air and Dust Sampling," Environmental Science and Technology, 14:47-50 (1980).

3. Hui, S. L. and J. O. Berger. "Empirical Bayes Estimation of Rates in Longitudinal Studies," J. of the Am. Stat. Assoc. 78:384:753-760 (1983).

4. Strenio, J. F., H. I. Weisberg, and A. S. Bryk. "Empirical Bayes Estimation of Individual Growth-curve Parameters and their Relationship to Covariates," Biometrics 39:71-86 (1983).

5. Mahaffey, K. R., J. L. Annest, H. E. Barbano, and R. S. Murphy. "Preliminary Analysis of Blood Lead Concentrations for Children and Adults: HANES II, 1976-1978," in Trace Substances in Environmental Health XIII, A Symposium, D. D. Hempill, Ed., pp. 37-51 (1979).

6. Walter, S. D., A. J. Yankel, and I. H. von Lindern. "Age-Specific Risk Factors for Lead Absorption in Children," Archives of Environmental Health 35:53-50 (1980).

7. Stark, A. D., R. Fitch Quah, J. W. Meigs, and R. Delouise. "Relationship of Sociodemogrpahic Factors to Blood Lead Concentrations in New Haven Children," J. of Epid. & Com. Hlth. 36:133-139 (1982).

8. Mahaffey, K. R. "Absorption of Lead by Infants and Young Children," Health Evaluation of Heavy Metals in Infant Formula and Junior Food, E. H. F. Schmidt and A. G. Hildebrandt, Eds. (Berlin: Springer-Verlag, 1983), pp. 69-85.

9. Charney, E., J. Sayre, and M. Coulter. "Increased Lead Absorption in Inner City Children: Where Does the Lead Come From?" Pediatrics 65:2:226-231 (1980).

CHAPTER 20

THE USE OF INDUSTRIAL HYGIENE DATA IN
OCCUPATIONAL EPIDEMIOLOGY

Robert F. Herrick and Larry J. Elliott

INTRODUCTION

The purpose of studies of occupational epidemiology is to
investigate the existence and nature of the associations between exposure to physical and chemical agents and outcomes
such as morbidity and mortality. While it is very difficult,
and some may claim it is impossible, to prove causality between
exposure and disease in the occupational setting, the validity
of associations observed in epidemiologic studies is determined
by estimating the probability that the observed associations
could be due to chance alone. The degree to which this causal
association can be established is, in part, determined by the
quality of the industrial hygiene data which are used to describe the exposure characteristics of the study population.
There are many factors and criteria which are used to assess
the validity of causal associations in epidemiologic studies;
Table 1 summarizes some commonly used examples [1]. Most of
the factors imply a measurement of exposure or dose, at least
qualitatively. This presentation discusses a model which may
be used to visualize the components of the exposure-response
relationship, with some examples of the use of industrial hygiene data in studies of occupational epidemiology. The model
is illustrated in Figure 1, and the first portion of the presentation will describe the components of the model, the measurement techniques applicable to each component, and the factors which mediate the pathways between the components.
    The second portion of the presentation discusses several
epidemiologic studies which use industrial hygiene data to help

Table 1.   Evaluation Criteria in Epidemiologic Studies.

---

o Strong   association   of   the   factor   (e.g.,   chemical
  exposure) to the outcome.

o A dose-response relationship between the factor and the
  outcome.

o A clear temporal relationship between the factor and the
  outcome.

o A biologically plausible explanation for the observed
  association.

o A consistency of findings across studies.

---

[a]Source: based on Lilienfeld, A. M., and D. E. Lilienfeld.
Foundations of Epidemiology (New York: Oxford University
Press, 1980).

characterize the exposure-response relationship. While the existence of a causal pathway (such as that described in Figure
1) is implicit in epidemiologic research, studies rarely characterize each separate component of the pathway. This is due,
for example, to the crudeness of the measurement techniques
available to assess dose in the occupational setting. Personal
exposure to a contaminant is usually the best surrogate measure
of dose which can possibly be made. Due to the retrospective
nature of most occupational epidemiology studies, even personal
exposures must often be estimated for workers who are no longer
employed, and may in fact be deceased. Actual historical measurements are often sparse, and of uncertain validity, while
changes in manufacturing processes make prediction of past exposures from present day measurements problematic. Despite
these and other limitations of methodology and data, the causal
associations described in the model have been successfully investigated, as the examples cited in the second portion of this
presentation illustrate.

THE CONCEPTUAL MODEL

Source

    Source may be defined as a point of emission, e.g., a coal-
fired power plant, a wastewater treatment facility, a gas sterilizer releasing ethylene oxide, a paint booth in an auto fac-

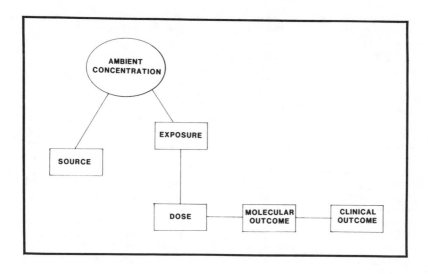

Figure 1.   Conceptual model of causal association in occupational epidemiology.

tory, a cyanide bath in a plating shop, a continuous mining machine in a coal mine.

The characteristics of the source are evaluated by methods such as stack sampling. The primary disadvantage of source sampling is that it does not include an assessment of human interaction with the source; however, good source monitoring is essential for effective control technology.

The first pathway is between the source and the ambient environment. Factors which mediate this pathway include size, location, and operating characteristics of the source, the nature of the emission, pollution controls, and the nature of the environmental matrix into which the contaminant is discharged.

Ambient Concentration

The ambient concentration is defined by the identification and quantitative determination of a chemical in an environmental matrix [2]. Examples typical of ambient concentration measurements made in occupational health are sulfur dioxide in air and benzene vapor in a control room of a refinery. Using this general definition of ambient concentration, lead dust on a cafeteria table in a foundry, or dioxin (TCDD) on a valve handle in a herbicide plant are also examples of ambient concentration.

Measurements of ambient concentrations are usually made by area sampling [2]. The techniques used to measure ambient concentrations in air usually involve locating a fixed monitor or collection system: this sampler provides a characterization of the area surrounding it. The primary advantage of this approach is that very sophisticated analytical instruments may be used in the occupational environment, e.g., portable gas chromatograph-mass spectrometer (GC-MS) systems. Large amounts of data may be collected to identify the components of complex mixtures, such as those found in many industrial solvents. These data may also be used to provide real-time measurements of concentration, which describe the emission pattern as well as the time-averaged concentration of a contaminant.

Although most area sampling methods apply to measuring airborne contaminants, methods have been developed for the measurement of surface contamination as well. Surface wiping techniques have been described in the Occupational Safety and Health Administration (OSHA), Industrial Hygiene Field Operations Manual [3], and extensive surface sampling data has been collected in plants producing herbicides and other products potentially contaminated with dioxin.

The disadvantage of these measurements is that they describe potential exposure only. Because most of these sophisticated area sampling devices are not truly portable, they cannot be used to directly sample the environment immediately surrounding the worker. In cases where workers spend their time at fixed work stations, these area samplers may provide good approximations of personal exposures, however, their lack of portability limits their usefulness for many industrial hygiene measurements. For example, infrared spectrophotometers are available for monitoring at fixed locations in the field, but the size and weight of these devices make it impractical to more than follow the activities of mobile workers.

The pathway between ambient concentration and exposure is in the focus of most industrial hygiene research. For the purposes of this discussion, exposure may be defined as ambient concentration modified by characteristics of both the contaminant and the individual in contact with it. These characteristics include:

o The physical and chemical properties of the contaminant
  For example, the size, shape, and density of an airborne particle determine how long a particle will remain in the atmosphere, whether or not it will be inhaled, and where it may be deposited in the respiratory tract. The solubility of a gas determines (in part) whether it will cause upper airway irritation or pulmonary edema. The boiling point of a solvent largely determines whether it is primarily hazardous to the skin via contact with the liquid or to the respiratory tract as a vapor. These few examples serve only to illustrate the wide range of physical and chemical properties which influence exposures to contaminants.

o The overall composition of the matrix
The characteristic most commonly measured is the concentration of a contaminant in its matrix, e.g., parts per million benzene in air; milligrams crystalline silica per cubic meter of air. Surface contamination may be described as the mass of contaminant on an area of known size, e.g., micrograms of lead per 100 cm^2. Other characteristics of the matrix can modify the pathway from ambient concentration to exposure, such as the presence of particulate material which can adsorb gases (including sulfur dioxide and formaldehyde) on its surface. In cases such as these, air sampling methods which measure contaminants in only one physical state (such as in the gaseous form) may underestimate the actual exposure.

o The pattern and duration of emission which produces the ambient concentration
For example, short but intense bursts of ethylene oxide (ETO) gas may escape from a sterilizer every time the door is opened. ETO gas is slowly released from sterilized products, resulting in a relatively constant, low level of ETO in the warehouse where sterilized products are stored [4]. Workers in these two exposure scenarios may have the same time-averaged exposure, but vastly different patterns of exposure. For substances which may produce health effects as a result of brief exposures, however, the significance of these peak exposures goes beyond their contribution to the time-averaged exposure level.

o Characteristics of the exposed individuals
Individual characteristics such as the amount of time spent in areas of high concentration; work activities resulting in high exposure (such as collection of a quality control sample from a reaction vessel); heavy work in a coal mine resulting in increased depth and rate of respiration; use of personal protective equipment; and personal characteristics such as smoking all modify the ambient concentration/exposure relationship [5].

Exposure

For this discussion, exposure is defined as ambient concentration modified by the factors described in the discussion of the pathway. Ambient concentrations may be manifested as exposures by multiple routes; lead is a good example [6]. Lead may be inhaled as very small particles of metal fume which are deposited deep in the lung; large particles of lead may be trapped in the upper respiratory tract and eventually swallowed; lead dust on hands may be ingested during smoking or eating; and lead in solution and organic lead compounds (such as tetraethyl lead) may be absorbed through the skin.

The industrial hygiene measurement used to characterize exposure to chemicals is the personal sample. Reflecting the emphasis on the respiratory route of exposure to chemicals, the most common type of personal sample is the breathing zone sample. The essence of this sampling method is that it seeks to define the personal exposure by collecting air from the microenvironment occupied by the worker. This simulation of human contact with a chemical is the primary advantage of the personal sample; these measurements are sometimes referred to as external dose. There is a great deal of interest in development of techniques to measure non-respiratory exposures, such as patches worn on the hand to measure the exposure by skin contact with compounds such as pesticides and aromatic amines, including methylene dianiline (MDA). This is an area of active research; however, the principles of dermal monitoring are summarized by Linch [7].

The disadvantage of these methods is that despite our efforts to define the nature and extent of exposure, these measures are still only surrogates of dose. In some cases they may be meaningful surrogates; in other cases, the measurements of exposure are made because they are the best we can do, but we really have little idea how well they describe an actual dose.

Dose

In defining the pathway between exposure and dose, we are leaving the world of the environmental scientist and entering that traditionally the province of the toxicologist and physician. Using the following operational definition of dose, we can link these disciplines by stating that dose is the physiologically significant component of exposure. Toxic effects are produced in a biological system when a chemical, or its metabolites or conversion products, reach the appropriate receptors in a system at a sufficient concentration and duration of contact to initiate a toxic manifestation [8]. Use of exposure measurements as correlates of dose implies, therefore, that we know something about the mechanisms of toxicity. This is possible for a few well-studied chemicals such as lead, carbon monoxide, and vinyl chloride. For most chemicals, however, the exposure-dose pathway is poorly defined [5]. We often use general models, based upon what is known about uptake, metabolism, and elimination of chemicals, to develop sampling strategies to evaluate the components of exposure which may be manifested as dose. Most personal exposure measurements are made, however, by applying the best available technique for measuring ambient concentration to the individual's working environment [2].

There are few documented techniques for estimating dose in an occupational setting. The techniques most commonly applied involve biological monitoring [9]. While biological monitoring is not itself a measure of actual dose, it is intended to improve the assessment of risk by measuring a parameter (the amount of a substance in the body) which is more closely related to the effect than is exposure (which is a measurement of the external environment). The best known biological monitoring technique is probably breath analysis for ethyl alcohol; drug screening in race horses and athletes is another common example. In the occupational setting, blood lead determination is probably the most common biological monitoring technique. The Occupational Safety and Health Administration (OSHA) lead standard includes the requirement for biological monitoring to assess lead exposure [10]. One great advantage of biological monitoring is that it allows measurement of chemicals which have entered the body by all routes, including inhalation, absorption through the skin, and ingestion through the gastrointestinal tract. The significance of these non-respiratory routes has been demonstrated in studies of workers exposed to metals such as lead [11]. Another advantage of biological monitoring is that it can reflect individual characteristics and work practices, e.g., the extent of skin contact and ingestion of a chemical, or the increased respiratory uptake due to physically demanding work. Another characteristic of biological monitoring is that the results may be influenced by non-occupational exposures [12]. In effect, the body serves as a 24 hour integrated sampler, reflecting occupational and non-occupational exposures.

There are several limitations to the use of biological monitoring as a complement to exposure measurements of the external environment (such as personal air samples). There are few thoroughly documented and validated methods for biological monitoring [9]. Frequently, the actual mechanism of toxic action is so poorly understood that the actual toxin is unknown, therefore, the compound which may be measured in the biological matrix may not be the proximate agent of toxicity. In other cases, the inaccessibility of the site of toxic effect limits our ability to directly sample the toxin. There is wide variability in the time course of substances in the body; some substances are rapidly eliminated, requiring that biological measurements be made almost immediately after exposure. For example, toluene exposure may be assessed by measuring hippuric acid levels in urine, but the biological half-life for this metabolite is one to two hours [13]. Other substances, such as lead and polychlorinated biphenyls, are slowly excreted from the body by a variety of routes including urinary, gastrointestinal, epithelial structures (such as hair), and sweat [12]. The time period between the appearance of a toxin at a receptor site, the initiation of a toxic effect, and the eventual expression of this toxicity as an observable outcome is usually not known. Despite these limitations, the development of biological monitoring techniques is proceeding rapidly. For example, the American Conference of Governmental Industrial Hygienists (ACGIH) has proposed six biological exposure indices as indicators of biological response to chemicals such as carbon monoxide, toluene, and xylenes [13].

The pathway from dose to molecular outcome is actually a gray zone; there are many factors which mediate this pathway, including variability in individual susceptibility to toxic effects. The distinguishing characteristic between dose and outcome is that the molecular outcome may be the first measurable effect on a living system, while all the previous elements in the continuum have described only the presence of a chemical in a variety of matrices. The dose-molecular outcome pathway is a major research area, and for purposes of this discussion it will be referred to as sort of a "black box" into which dose enters and an observable effect may emerge.

Molecular Outcome

The molecular outcome is defined as the earliest observable effect; this effect may be the alkylation of genetic material, an increase in the rate of sister chromatid exchange, the inhibition of an enzyme system, or the development of an immune response. This definition is subject to constant revision as our ability to detect toxic outcomes due to chemical exposures improves. The greatest advantage of measuring molecular outcomes is that they are conclusive evidence that the elements described so far (source, ambient concentration, exposure, dose) have led to a response; this is at the same time a great disadvantage because this measurement is no longer just a predicator of risk; it is evidence of a response, and a potentially toxic effect, at the molecular level.

Clinical Outcome

The distinction between a molecular and a clinical outcome is intended to differentiate toxic outcomes which are observable only by cytological or biochemical techniques from those which may be observed by measuring outcomes such as death, illness, or impaired function. The latter are the classic outcome measures of occupational epidemiology.

Summary

The conceptual model links the components of the causal pathway from the point at which a contaminant is released to the environment to the manifestation of a health effect. Occupational health research attempts to characterize the components of this model, and the factors which mediate the path-

ways between these components. In testing the validity of
causal associations observed between the components, the evaluation criteria in Table 1 are applied [1]. While a study is
not required to satisfy all these criteria to define the relationship between exposure and outcome, the quality of the exposure assessments has a major impact on the ability of a study
to satisfy these criteria. Some examples will be presented in
the following section to illustrate the use of industrial hygiene data to meet the criteria for establishing causality in
epidemiological studies.

EXPOSURE ASSESSMENTS IN OCCUPATIONAL EPIDEMIOLOGY

    The methods of occupational epidemiology may be classified
into three general study types, based upon the point in time at
which observations of exposure and outcome are made. The study
types are retrospective, prospective, and cross-sectional
[14]. In the retrospective cohort study, a population is classified on the basis of its exposure after disease or death has
occurred. The morbidity or mortality experience is compared
between the exposed cohort and some referent population, such
as the general population. Another type of retrospective study
used in occupational epidemiology is the case control study, in
which the study population is divided on the basis of the presence or absence of disease; one looks backward from outcome to
exposure, testing the association between exposure and disease. In studies of occupationally exposed populations, case
control studies are often done after a retrospective cohort
study has been completed; these are described as nested case
control studies. In prospective cohort studies, the study
groups are once again classified on the basis of exposure and
are followed forward through time to observe the development of
outcomes such as illness and death. The final type of study is
a cross-sectional study, in which persons are selected for inclusion in the study at a point in time, without regard to
their previous exposure or disease status; then exposure and
disease are determined at the same time. The inherent weakness
in cross-sectional studies is that they do not allow evaluation
of the exposure/disease time sequence. The application of exposure assessments to epidemiologic research is discussed in
the following section.

Retrospective Studies

    In order to fulfill the criteria for establishing a causal
relationship between exposure and outcome in retrospective
studies, it is necessary to reconstruct historical exposures

(to the extent possible) as they existed during the work years of the study population. This is often a very difficult task, due to lack of historical exposure measurements, incomplete work histories, and changes in manufacturing processes, control technology, and industrial hygiene measurement techniques. The quality of information available from these sources determines the extent to which the study population can be divided into groups which reflect their exposure histories. Lack of definitive historical exposure classifications is unlikely to result in the incorrect association of exposure and outcome when one, in fact, exists. Inaccurate or incomplete exposure classification is, in most cases, more likely to result in misclassification, such as the incorrect assignment of highly exposed workers to a low exposure group, or the reverse. If this misclassification is random, as would be expected when it results from incomplete exposure data, the errors will obscure the true exposure-effect relationships and create a bias toward negative conclusions.

In the absence of historical exposure information, the simplest approach to exposure estimation is to use duration of either employment or exposure as a surrogate of dose. The disadvantages of this approach are many, one of which is that duration is often a poor surrogate of exposure (and therefore dose), obscuring the true dose-outcome relationship. This approach has been used in preliminary analysis to divide cohorts on the basis of duration of employment as a surrogate of cumulative exposure. In early studies of rubber workers exposed to benzene, an association was observed between total years of benzene exposure and risk of leukemia, even though the atmospheric benzene concentrations were not known [15]. Angiosarcoma of the liver was observed primarily among workers with more than 10 years of exposure to vinyl chloride in cleaning of reaction vessels [16].

The next level of sophistication in exposure assessment is the assignment of cohort members to qualitative categories based upon ranking of the magnitude of their exposures. For example, an indicator of relative exposure can be selected and the cohort divided into categories which reflect their exposure ranking. Nature of exposure, i.e., direct or indirect, has been used to categorize workers. Job title may also be a useful indicator of exposure, allowing workers to be ranked on the basis of their job histories. For example, in a study of workers exposed to sulfuric acid mist in steel pickling operations, a group of workers known to have been exposed to high levels of sulfuric acid mist was identified by examining job histories, historical industrial hygiene and engineering records, and by observations made on walk-through surveys. These workers were compared with workers exposed to mixtures of acids, those exposed to any level of sulfuric acid, and those never exposed to sulfuric acid. Death rates for these groups were compared, and all acid-exposed workers were found to be at excess risk of dying of lung cancer; however, this excess was not statistically significant [17]. Workers exposed to high levels of sulfuric acid were found to be at greater risk than those exposed to sulfuric acid at any level. While findings such as these are certainly suggestive, they do not provide the sort of information needed to accurately define the exposure/response relationship.

If sufficient historical exposure information exists, or can be derived, exposure values which are characteristic of each job assignment or task can be used to develop semi-quantitative exposure classifications. This approach has been used in studies of workers exposed to asbestos and benzene.

In the case of asbestos, a retrospective mortality study was conducted in a plant which processed chrysotile into asbestos textile from 1896 to 1975 [18,19]. Airborne asbestos fiber concentrations had been measured by the company, an insurance carrier, and the U.S. Public Health Service, from 1930 to 1975. By using the approximately 6,000 air sampling measurements, detailed process descriptions, and documented changes in the manufacturing processes and control technologies (primarily ventilation), an exposure classification model was developed. The model was constructed by dividing the factory into eight exposure zones, and classifying jobs within each zone into Uniform Job Categories. The effect of a number of variables on asbestos exposures was considered, and improvements in ventilation and changes in production volume were found to be significantly associated with exposures. These factors were included in a multiple regression model, resulting in a series of predictive equations of the following form:

$$\bar{Y}_i = \sum_k \beta_{ik} Z_{ik} + \sum_j \alpha_{ij} Z_{ij} + \sum_t \delta_{it} Z_{it}$$

where:

$\bar{Y}_i$ is the mean log asbestos concentration for exposure zone i

β_{ik} is the multiple regression parameter for job k in zone i

α_{ij} is the multiple regression parameter for control j in exposure zone i

δ_{it} is the multiple regression parameter for time interval t in exposure zone i

Z_{ik} (or j or t) is an independent variable (0 or 1) used to identify job k (or control j, or time period t) in exposure zone i

Using the available historical measurements of airborne asbestos concentration, each industrial hygiene sample was assigned to an exposure zone and Uniform Job Category. By including the variables for engineering controls and time period

for each sample, the model parameters β, α, and δ were estimated by least squares fitting procedures. The significance
level for each model parameter was tested, and the model was
then used to predict mean exposure levels with 95% confidence
intervals for each job where actual historical measurements
were not available. These predicted values compared well with
historical exposure measurements made in similar asbestos processing facilities. By combining the predicted exposure values
with the detailed occupational histories, a job-exposure matrix
was constructed, and cumulative exposures were used to stratify
the cohort into categories as shown in Table 2.

Table 2. Exposure-Response Relationships for Lung Cancer Among
Chrysotile-Exposed White Males With At Least 15 Years
Latency.[a]

Cumulative Exposure Fiber/cm^3 x Days	Lung Cancer (ICDA[b] 162,163)		
	Observed	Expected	SMR[c]
1,000	5	3.58	140
1,000-10,000	9	3.23	279
10,000-40,000	7	1.99	352
40,000-100,000	10	0.91	1099
>100,000	2	0.11	1818
Overall	33	9.82	336

[a]Source: Dement, J. M., R. L. Harris, M. J. Symons, and C.
M. Shy. "Exposures and Mortality Among Chrysotile Asbestos
Workers I. Exposure Estimates," Am. J. Ind. Med. 4:399-419
(1983).
[b]International List of Diseases and Causes of Death.
[c]SMR = Standardized Mortality Ratio.

A recent study of rubber workers exposed to benzene illustrates the use of historical exposure measurements to reconstruct exposure histories for cohort mortality analysis, followed by a case control study of the same worker population
[20]. Prior studies had shown excess leukemia in this study
population, and this study was undertaken to quantify the
exposure-effect relationship. For each worker potentially exposed to benzene, the worker's department and his actual work
activities were determined, and his job title was assigned a
numeric code. The codes were then fitted into exposure classes. These classes corresponded to areas where industrial hygiene data had been collected. Job-exposure matrices, which

tabulated job classes by year, were constructed, and the available industrial hygiene measurements were entered into their
cells in the matrix. Using information available on manufacturing process changes, addition of control technology, and the
available air sampling data, a set of rules for interpolation
between the known data points was developed. Cells for which
there was no measurement data available were filled according
to these rules.

For each member of the study population, cumulative lifetime benzene exposures were calculated by summing the daily
predicted exposure values over each individual's working lifetime. The cohort was then divided into four exposure strata,
as shown in Table 3. The boundaries of these strata correspond
to the cumulative lifetime exposures which would be accumulated
by workers spending a 40 year working career in atmospheres of
less than 1-5, 5-10, and greater than 10 ppm benzene.

Table 3. Observed and Expected Deaths from Leukemia in Benzene
Exposed Workers.[a]

Deaths	Cumulative Exposure (PPM-years)				Total
	<40	40-200	200-400	>400	
Observed	2	2	3	2	9
Expected	1.88	0.60	0.21	0.05	2.74
Standardized mortality ratio	106	334	1444	3883	328
(CI[b])	(12-384)	(38-1207)	(290-4220)	(436-14201)	(150-623)

[a]Source: Rinsky, R. A., A. B. Smith, R. Hornung, T. G.
Filloon, R. J. Young, A. H. Okun, and P. J. Landrigan.
"Benzene and Leukemia: An Epidemiologic Risk Assessment,"
(in press).
[b]95% Confidence Interval.

In addition to the standardized mortality ratio analysis
just described, a matched case control analysis was also performed. Conditional logistic regression was used to compare
exposure histories of workers known to have died of leukemia
with controls, who were workers known to have died of other
causes. Using the exposure estimates previously developed, the

cases and controls were compared by their cumulative (lifetime) benzene exposure, duration of exposure, and exposure rate, which was calculated by dividing cumulative exposure by duration of exposure. Logistic regression models of the form

$$OR = \exp(B_1 X_1 + B_2 X_2 + \ldots\ldots + B_n X_n)$$

were used, where OR is the odds ratio, which is approximately the relative risk of dying of leukemia among the exposed group, divided by the relative risk of dying of leukemia among the unexposed group. The X terms correspond to the exposure variables being tested, which were cumulative exposure, duration of exposure and exposure rate, and the B terms were the regression coefficients which were estimated using the model. By testing a number of models which included these exposure variables singly and in combination, cumulative exposure was found to be the best predictor of death from leukemia. The best fitting model to describe the odds ratio for leukemia in relation to cumulative exposure to benzene was

$$OR = \exp (0.0135 \times \text{ppm-years}).$$

This study illustrates the use of maximum likelihood estimates to reconstruct historical exposures, and the use of this reconstruction in analysis of the exposure-outcome relationship.

Occasionally there is sufficient personal exposure information available to allow individual exposure measurements to be used in reconstructing exposure histories for each member of the study population. For example, in a study of workers exposed to ionizing radiation at a naval shipyard, personal monitoring data was available in the form of radiation film badges and dosimeters for all workers potentially exposed to radiation [21]. Using this personal exposure information, cumulative lifetime exposures were calculated for each worker, and the population was divided into exposure categories. The mortality experience of the workers in these categories was analyzed by several methods, and no associations between radiation exposures and excess mortality were observed, as shown in Table 4.

Prospective Cohort Studies

In prospective cohort studies, the study population is divided on the basis of exposure category, then followed through time to measure the incidence of outcomes. Several large, population-based studies have been performed to study risk factors associated with heart disease and to follow cigarette smokers over time, but few prospective studies have been undertaken in the occupational setting. Cost and the amount of time required to conduct a prospective study of a disease with long latency make prospective studies uncommon in occupational

Table 4.   Deaths for all Malignant Neoplasms by Cumulative
Radiation Exposure Among Shipyard Workers.[a]

Cumulative Radiation Dose (rem)	Observed	Expected	SMR[b]
0.001 - 0.029	29	33.2	87.4
0.030 - 0.009	32	37.1	86.3
0.100 - 0.499	46	56.8	81.0
0.500 - 0.999	26	23.5	111.0
1.00 - 4.99	45	42.8	105.0
5.00 - 14.99	17	18.0	94.4
15.000 and over	6	7.2	83.3
Total	201	218.5	92.0

[a]Source: Rinsky, R. A. et al.   "Cancer Mortality at a Naval
Nuclear Shipyard," Lancet, 1:231-235 (1981).
[b]Standardized Mortality Ratio.

health research.   With the advent of medical screening and
health surveillance programs, however, prospective studies are
becoming more attractive as ways of performing comprehensive
studies of exposure and outcome.   Several companies have developed computer-based occupational health and environmental surveillance systems to prospectively monitor employee health
status [22].

Cross-Sectional Studies

Cross-sectional studies also use occupational exposure assessments.   These studies measure the prevalence of disease
among active workers at the same time exposure status is determined.   The cross-sectional study design does not allow examination of the temporal relationship between exposure and disease, and is not well suited for study of diseases such as
cancer, which have a long period of latency between the time of
exposure and expression of the disease.   However, for studies
of morbidity, such as pulmonary or reproductive function, the
cross-sectional study can be very useful.
An example of such a study is an investigation of the possible assocation between fluorocarbon exposures and cardiac ar-
rythmias.   A population of workers using the fluorocarbon Freon
113 has been identified.   Freon 113 is being used as a solvent
to clean metal parts.   In a study currently being designed, ex-

posures will be assessed, using personal samplers and a portable infrared analyzer to measure the high peak exposures which
correspond to job tasks requiring direct contact with the solvent. Each worker will also wear a monitor which will continuously record his electrocardiogram (ECG). By comparing the individual patterns of exposure to Freon and the outcome as recorded by the continuous ECG patterns, the potential association between Freon exposure and cardiac arrythmia can be evaluated.

CONCLUSIONS

    As the level of sophistication of epidemiologic research
rises, the need for complete, accurate assessments of exposure
has become apparent. In fact, the ability of epidemiological
studies to define the true association between disease and
workplace exposures is often limited by the quality of the exposure assessments available. The conceptual model described
in this paper can serve as a useful framework for describing
the relationships between the elements of the causal pathway
which is the subject of epidemiologic research. The examples
cited represent early efforts in the process of developing the
research methodologies needed to explore the relationships between occupational exposures and disease.

DISCLAIMER

    The work described in this chapter was not funded by EPA
and no official endorsement should be inferred.

REFERENCES

1. Lilienfeld, A. M., and D. E. Lilienfeld. Foundations of
    Epidemiology (New York: Oxford University Press, 1980).

2. Hosey, A. H. "General Principles in Evaluating the Occu-
    pational Environment," in The Industrial Environment - Its
    Evaluation and Control, U.S. Department of Health, Edu-
    cation and Welfare, NIOSH, 1973, p. 98.

3. "OSHA Industrial Hygiene Field Operations Manual," United States Department of Labor, Occupational Safety and Health Administration (1979).

4. Roy, P. A. "Engineering Control of Ethylene Oxide Exposures from Gas Sterilization," in The Safe Use of Ethylene Oxide: Proceedings of the Educational Seminar (Washington: Health Industry Manufacturers Association, 1981) pp. 193-208.

5. McClellan, R. O. "Health Effects of Diesel Exhaust: A Case Study in Risk Assessment," Am. Ind. Hyg. Assoc. J. 47(1):1-13 (1986).

6. "Criteria for a Recommended Standard...Occupational Exposure to Inorganic Lead, Revised Criteria - 1978," DHEW (NIOSH) Publication No. 78-158 (1978) pp. III 1-18.

7. Linch, A. L. Biological Monitoring for Industrial Chemical Exposure Control (Cleveland, OH: CRC Press, 1974), Chapter V.

8. Doull, J. "Factors Influencing Toxicology," in Casarett and Doull's Toxicology, J. Doull, C. D. Klaassen, and M. O. Amdur, Eds. (New York: Macmillan Publishing Co., Inc., 1980), p. 70.

9. Lauwerys, R. R. Industrial Chemical Exposure: Guidelines for Biological Monitoring, (Davis, CA: Biomedical Publication, 1983), Chapter I.

10. Federal Register, November 14, 1978, Vol. 43, p. 52952.

11. Kehoe, R. A. "Standards for the Prevention of Occupational Lead Poisoning," Arch. Environ. Health 23:245-248 (1971).

12. Hernberg, S. "Lead," in Biological Monitoring and Surveillance of Workers Exposed to Chemicals, A. Aitio, V. Riihimaki, and H. Vainio, Eds. (Washington: Hemisphere Publishing Corporation, 1984), p. 20.

13. American Conference of Governmental Industrial Hygienists. Documentation of the Threshold Limit Values, Fourth Edition. (Cincinnati, OH: ACGIH, Inc., 1980), p. BEI-19(84).

14. Monson, R. R. Occupational Epidemiology (Boca Raton, FL: CRC Press, Inc., 1980), Chapter 3.

15. Infante, D. F., R. A. Rinsky, J. K. Wagoner, and R. J. Young. "Leukemia in Benzene Workers," Lancet 2:76-78 (1977).

16. Tabershaw, I. R., and W. R. Gaffey. "Mortality Study of Workers in the Manufacture of Vinyl Chloride and its Polymers," J. Occup. Med. 16:509 (1974).

17. Beaumont, J. J. et al. "Mortality Among Workers Exposed to Sulfuric Acid Mist in Steel Pickling Operations," In press.

18. Dement, J. M., R. L. Harris, M. J. Symons, and C. M. Shy. "Exposures and Mortality Among Chrysotile Asbestos Workers I. Exposure Estimates," Am. J. Ind. Med. 4:399-419 (1983).

19. Dement, J. M., R. L. Harris, M. J. Symons, and C. M. Shy. "Exposures and Mortality Among Chrysotile Asbestos Workers II. Mortality," Am. J. Ind. Med. 4:421-433 (1983).

20. Rinsky, R. A., A. B. Smith, R. Hornung, T. G. Filloon, R. J. Young, A. H. Okun, and P. J. Landrigan. "Benzene and Leukemia: An Epidemiologic Risk Assessment," In press.

21. Rinsky, R. A. et al. "Cancer Mortality at a Naval Nuclear Shipyard," Lancet 1:231-235 (1981).

22. Lynch, J. "Industrial Hygiene Records - Will They be Useful," Ann. Am. Conf. Gov. Ind. Hyg. Vol. 6:67-73 (1983).

INDEX

Acceptable Daily Intakes (ADI), 141, 144
Aerometric and Emissions Reporting System (AEROS), 149
 Compliance Data System (CDS), 152
 Comprehensive Data Handling System (CDHS), 150
 Hazardous and Trace Emissions System (HATREMS), 150
 National Emissions Data System (NEDS), 149-152
 Quality Assurance Management Information System (QAMIS), 150
 Source Test Data System (SOTDAT), 150
 Storage and Retrieval of Aerometric Data System (SAROAD),
  149-152
aflatoxin B_1 (AFB)
 bonding with DNA, 2, 8-9
 carcinogenic effects, 8-9
 human exposure, 1-2
 hydrolytic pathways, 9
 in food, 10
 metabolites, 2
aflatoxin B_1-guanine (AFB-Gua)
 HPLC profile, 2-4
 isolation, 3
 synchronous fluorescence spectrum, 5-8
aflatoxin B_1-diol (AFBdiol), 9
air data bases
 National Air Data Branch (NADB), 149
air, indoor, See indoor air
albumin, 23
alkylating carcinogens
 binding effects of exposure, 29-30
 derivatization and GC-MS, 30-31
 hemoglobin alkylation and exposure monitoring, 31-35
 isolation of alkylated amino acids and purines, 30
 monitoring human exposure, 29-30, 36-37
ambient concentration, 261-263
American Cancer Society, 127
American Chemical Society, 127
American Conference of Governmental Industrial Hygienists
  (ACGIH), 265
American Hospital Association, 129
American Water Works Association (AWWA), 135, 139
aromatic amines
 analysis for cysteine sulfinamides, 61-62
 environmental occurrence, 58-59
 hemoglobin levels in smokers and nonsmokers, 62-64
 metabolic activation, 59-60
 proposed mechanism of carcinogenicity, 64